Solido Toys

Dr. Edward Force

Schiffer Publishing Ltd

77 Lower Valley Road, Atglen, PA 19310

Variation & Price Guide

SOLIDO
With Price Guide and Variations List

Dr. Edward Force

Schiffer Publishing Ltd.

The names of **Solido** and its various series of miniature vehicles, of the real vehicles of which they are models, and of the firms and products that appear in advertising on the models, are registered trade marks.

Printed in the United States of America,

Library of Congress Number: 93-85083
ISBN: 0-88740-532-0

Published by Schiffer Publishing, Ltd.
77 Lower Valley Road
Atglen, PA 19310
Please write for a free catalog.
This book may be purchased from the publisher.
Please include $2.95 postage.
Try your bookstore first.

We are interested in hearing from authors
with book ideas on related subjects.

Contents

SOLIDO

The year was 1930. Ferdinand de Vazeilles of Nanterre, France had been asked by the makers of Gergovia spark plugs to produce a promotional item to advertise their product. His design of a spark plug on wheels, complete with fenders and running boards, was manufactured by a local foundry. The first picture I saw of one made me wonder what on earth it was supposed to be, but subsequent pictures from various angles showed that this funny little vehicle does indeed represent a spark plug. Regardless of its looks, it sparked several series of miniature vehicles in the 1930s, produced by Ferdinand de Vazeilles under the name of Solido, the firm he founded in 1932. The 1/35 scale Major series was born in 1932, the 1/40 Junior series in 1933 and the 1/50 Baby series in 1935.

All of these models followed a basic principle of early Solido workmanship: All the body and interior parts were made to fit a single chassis used by the entire series. In some cases, the Major models used several body components, but the standard procedure for the smaller scales was to use a single body casting whenever possible. The bodies or body parts, seats, and other components, including clockwork motors, could all be bolted to the chassis. The models were sometimes sold in kit form, and the owner could easily switch bodies on chassis, install or remove motors and, in the case of models with several body components, create a variety of vehicles to his own taste.

World War II naturally interrupted the production of Solido miniatures, but that production resumed afterward, though the big Major models had already been phased out in 1937. I am not sure why, but it is possible that the expense of casting multiple body parts made them economically unprofitable. Some of them have suffered from what is known in the vulgate as metal fatigue, and the surviving examples, whose body styles are more easily categorized than their radiator grilles, are rare and fragile relics of Solido's early years.

The Junior and Baby series remained in production until the mid-Fifties, and they were joined in 1952 by twelve 1/60 scale models that formed the Mosquito series. No additions were ever made to this series, and in fact the three series of vehicles all looked primitive compared to the other brands of miniature autos that were coming on the market at the time. For the Fifties

were France's Golden Age of miniature cars, and the history of the other brands that made it a wonderful time for collectors is dealt with in this author's *Classic Miniature Vehicles: Made in France*. The Solido management must have realized that new and better products, built to a new and high standard of quality, were needed to assure the firm's health and reputation.

It was this new 1/43 scale 100 series, which first appeared on the market in 1957, that made Solido a classic manufacturer of miniature vehicles. Their quality was excellent from the start and grew even better over the years, and they included the latest features found in other models on the market at that time: spring suspension, clear plastic windows and windshields, interior detail and even driver figures--ranging from helmeted racing drivers to young ladies in low-cut dresses--figures indeed!

These new Solido models quickly established themselves on the market as favorites among children and collectors alike, but it was only in 1962 that the firm realized that life would be less complex if they numbered their products. The first model issued in 1957, the Le Mans D-Type Jaguar, was given number 100, and subsequent models were numbered in chronological order up to 124, the Abarth 1000, which was the first member of the series to see the light of day already equipped with a catalog number. Somewhere along the way, though, a Ford Thunderbird seems to have been missed, and it was given the slightly anomalous number of 113b, though it bore no relationship to the #113 Fiat-Abarth speed record car. In fact, the Thunderbird bore a certain relationship to the very last pre-100 series Junior models; both it and the 115 Rolls-Royce were based on Junior series models, and the #119 and 120 Chausson buses also showed similarities to earlier model construction.

By the time the numbering system was instituted, two other series of Solido models had been created. The first commercial vehicles, which became the 300 series, came on the market in 1960, and the first military models of the 200 series appeared in 1961. New series were created later as new types of models were produced. In 1970 the 600 series of large-scale cars and trailers appeared; they were not very popular, and subsequent numbers in the 600 series were given to sets of 1/43 scale vehicles. A small 500 series of farm tractors and trailers came on the market in 1977, and in 1980 an even smaller 400 series of two buses was about to appear when the firm changed hands and numbers.

But we are getting ahead of ourselves, for in 1972 number 199 was used. The 100-199 series included many racing, sports and touring cars, a few cars with trailers, and several superb "Age d'Or" (Golden Age) models of classic cars of prewar days. It is unfortunate that it did not occur to Solido at the time to give these classic models a numbering series of their own--an error that had to be rectified in more modern times.

But with the 100 series numbers used up and more car models on the way, new numbers had to be found, and in 1973 the 10 series was born. It too included classic as well as modern cars, and as 1980 dawned it had reached number 97. What next?

The year of 1980 brought a much greater change than that of catalog numbers, though it introduced a revolution in them as well. The Solido firm, like others in the field, was having its financial problems, and in 1980 the company was bought by the manufacturers of Majorette miniatures. Majorette might well be described as the French equivalent of Matchbox, for its specialty was pocket-size models ranging in size from about 1/50 to 1/100 scale. The union of this firm with the basically 1/43 scale Solido models (though it must be added that many of the trucks, tanks and other large models were made to 1/50, 1/60 and other scales) was a sensible one, as it covered two of the most popular basic scales or sizes of diecast miniature vehicles.

Solido's new ownership breathed new life into the firm without harming its tradition of superb quality--for by now Solido was regarded as one of the world's finest brands of diecast miniatures. The quality remained excellent, distribution seems to have improved, and prices certainly came down--good news indeed to devoted Solido collectors all over the world. In addition, the new management tried to reorganize Solido's catalog numbering system, and it must be said that reorganization was needed.

The numbering system introduced in 1980 consisted of four digits, which gave much more room for expansion, and included the following series:

1000: Touring and sports cars of the 10 series, plus new additions.
1100: A few survivors of the 100 series; they did not last long.
1300: A new series of touring and sports cars, originally called Cougar and also marketed by Dinky Toys.
2200: Military vehicles, basically the 200 series.
3300: Commercial and emergency vehicles from the 300 series.
3500: The farm vehicles, beginning with #3510, plus a #3650 set.
3800: Helicopters, beginning with the few from the 300 series.
4000: Classic oldtimers from the 10 series, plus new additions.
4100: Classic oldtimers from the 100 series.
4400: The two buses and other vintage commercial vehicles.
5000: Kits, formerly using the number of the readymade model followed by the letter K.
6300: Circus models, formerly part of the 300 series.
6600: Circus sets, formerly part of the 600 series.
7600: Sets, formerly the 610 series.

This system looked good on paper, but the management found it practical to make changes in 1982-83 and again in 1985-86. Several new series were introduced as well, and the post-1980 series include the following:

1200: Versions of 1300 series models, made in Portugal.
1500: Classic sports cars, introduced in 1988.
1700: Special versions of sports cars, introduced at the end of 1980 and later turned over to the Top 43 firm.
1800: Reissues of models from the seventies.

2000: A few emergency and commercial vehicles, some later moved to the 2100 series. These numbers were introduced in 1983 and the last of them were phased out in 1988.

2100: A newer series of emergency and public service vehicles, introduced in 1986 and still in use. It includes a few models that previously had 300, 3300 and 2000 numbers.

3000: A small series of two-piece items: cars or trucks and trailers, or in one case a wrecker pulling a car. This series was phased out by 1988.

3100: Larger emergency and public service vehicles, including a few survivors of the 300, 3000 and 3300 series; like the 2100 series, it is now used for fire vehicles.

3500: When the farm vehicles were dropped, 3500 numbers were given to semi-trailers and other large commercial vehicles.

3700: One auto transporter.

4500: A new series of classic postwar cars, born in 1984.

6000: A new series of military vehicles, introduced in 1986.

7000: Various sets, plus a new version of the #3700 car carrier and a low loader resurrected from the 300 series.

8000: A new Prestige Series of large-scale (about 1/18) cars and trucks, introduced in 1987.

9000: Promotional models.

This system has room for everything, establishes a more or less ascending order of prices (though some of the 4000 models may cost less than some of the 3000's), and allows for future expansion, so we can hope that, though new series will doubtless be introduced in the future, there will be no more need to renumber existing models. The financial stability of a firm that now includes Norev as well as Solido and Majorette is rumored to be questionable as I write this (January 1993), but let us hope that Solido models will survive for many more years and continue to rank as a truly classic brand of diecast miniature vehicles, a brand that attracts new admirers while it continues to fascinate those of us who discovered it many years ago.

But this alone does not tell the whole story of Solido, for the firm has worked with other firms, both inside and outside France, at various times in its history. Beginning in the Fifties, the Spanish firm of Dalia produced Solido models under license for the Spanish market. A Solido branch in Brazil, Buby of Argentina, Tekno of Denmark, Lyra of Greece, the Louis Marx branch in Hong Kong, and someone in Japan all produced Solido or Solido-based models. In 1980 Solido collaborated with another classic firm that was suffering financially: the makers of Dinky Toys, to produce a small series of Cougar models originally numbered in a separate 1300-1400 series as Solido models and later absorbed into the Solido 1300 series. And in more recent years Solido castings, or models with painted bodies but without decals, have been supplied to such firms as Cofradis, Verem and the aforementioned Top 43 for adaptation and sale, usually as sports cars with decals or labels applied to represent actual competition cars, or with similar additions to serve as emergency vehicles. The market for emergency vehicles, and in particular for fire vehicles, has grown tremendously, and Solido itself and these related firms have recognized this fact and put a great number and variety of such models on the market.

In addition to all this, numerous Solido models have been used for promotional purposes by collectors' clubs, model car dealers, commercial firms and even the U.S. Marine Corps. It is up to the individual collector to decide which of these promotional models, if any, to collect–a decision that is often made easier when one looks at the prices charged for some of them. Some of these models have been wholly produced by Solido (Code 1), others have been adapted by others with the firm's permission (Code 2), and there are probably some Code 3 specimens in existence as well--models with additions not known to or authorized by Solido--but my data do not always tell me which are which. Along with this, models that came in Top 43, Cofradis or other firms' boxes can be removed from these boxes and appear to be ordinary Solido models, and it is possible that a few models listed in this book were not issued by Solido in the colors in which they now appear.

As for what the listed models are worth, this book will include a price guide which will try to set a pair of values, high and low, for each model or group of models. As usual in this series of books, the high value for a mint boxed model will be that which a collector who simply must have that model might be willing to pay; the lower price, also for a mint boxed specimen, will represent what the garden-variety collector might be willing to pay for a pleasant model. A mint but unboxed model might sell for about 10% less than the price in the price guide, while non-mint models would be worth considerably less, depending on their rarity and condition.

Please do not regard this price guide as universal law! As anyone who has attended a toy show knows, the same model can be worth different prices to different people. We can only state values that seem appropriate to this author at the time of writing this book, for the purpose of giving the reader a guide, a rough idea of the models' values, so please do not regard the price guide as having the strength of international law yourself or expect anyone else to regard it in that light.

This book could not have been written without the pioneering work of other scholars that preceded it. Much information has come from the yearly catalogs published by Solido, as well as from the *Catalogue Mondial des Modeles Reduits Automobiles*, compiled by Jacques Greilsamer and Bertrand Azema and published in 1967, from Volume 1, No. 1 to 4 of the Club Solido journal, which lists everything issued from 1986 to 1988, including promotional models, and from Bertrand

Azema's superb and minutely detailed *Solido, Catalogue d'un univers*, Volume 1, covering pre-1957 and listing post-1982 models, and Volume 2, covering models from 1957 to 1982. The serious collector of Solido models will find these expensive large-format volumes well worth the price, as they include a truly incredible wealth of information on models, colors, castings and the like, in far greater detail than could or should be included in this book, in which only major variations will be listed.

The descriptions of the models will not include every single detail, but will try to strike a happy medium between brevity and thoroughness. The cast metal parts will be noted first, then the plastic parts, usually beginning with the windows or windshield (if no color is noted, then the plastic is colorless), then the running gear ("hubs" and "tires" are two separate entities; "wheels" include both in one piece, though they may have hub inserts added to them). There was a time when Solido made hubs and tires of plastics that interacted with each other, with dreadful results. This affected some of the large trucks, as well as cars such as the Fiat X1/9.

The last part of the description will include painted trim, decals or labels. A few words about decals–I HATE THEM!!! I've tried to put them on the models that came with them, but when I've run out of cuss-words and the decals are too small to handle, I give up. Sometimes I'm not even sure where they are supposed to go. So don't take the number or arrangement of the decals in the photos for gospel.

Major color variations will be listed in the following general order: gold, silver, white, cream, red, orange, yellow, green, blue, purple, tan, brown, gray, black. Light and dark shades will be noted, and just plain "green" or "blue", for instance, will mean a medium shade. Please bear in mind that I have not seen every color variation, and that some listings are based on those in the Club Solido journals or M. Azema's books. Being French, he can describe colors in terms of French wines–or chestnuts–and know that French collectors, or even highly civilized other Europeans, will understand. Alas, I am not only a small-town American, but also, as they say in Germany, 100% alcohol-free, and I haven't decided whether to record "Bordeaux" as "wine red" or "maroon"; you may find both terms in this book. "Marron", on the other hand, is the French word for "chestnut" and appears to be a darker shade of tan than "sable" (sand color). As for "marron glacée", what color is a glazed chestnut?

For reasons of space, neither casting variations nor minor color or logo variations will be listed. There are numerous minor casting variations in Solido models, and the serious collector will find a wealth of detail in M. Azema's books, as well as lists of accessories, sets of components, and non-automotive products. And what with Solido's practice of

putting one decal on a model and packaging the rest with it, an endless variety is not only possible but invited!

In addition to M. Azema's books, numerous collectors' journals have been of help in tracing the history of Solido, and many collectors and dealers have been of help in obtaining the models. I am thankful that I collected Solido models assiduously up to a few years ago, and though I had to devote my time and money to other models and other pursuits for a few years, I am happy to return to serious collecting of Solido models. A single visit to James Wieland's basement added well over 100 Solido models to my collection, and subsequent purchases from various dealers have also helped to fill holes in my collection, which numbered over 1200 Solido models by the time this book was finished. I still lack a number of models, and would be happy to obtain them if you have them to spare. Any additional information on the models would also be most welcome.

Hearty thanks to all the collectors, dealers and scholars who have contributed to the compilation of this chronicle, to Peter Schiffer and his staff–particularly photographer-editor Doug Congdon-Martin–for making its publication possible, and as always to Pat, Lisa and Ellen for allowing me the time, energy and peace of mind–though this last commodity is often questionable–to write it. Thanks to you as well for buying and reading it, and I hope you enjoy collecting Solido models every bit as much as I do!

Dr. Edward Force
42 Warham Street
Windsor CT 06095

Row 1: **Junior 56 Coupé de Ville, 57 Conduite Intérieure, 64 Roadster (x2).**
Row 2: **59 Fourgon, 61 Camion Citerne (x2), 62 Autocar.**
Row 3: **60 Camion (= crane of #74), 66 Autocar, 67 Berline Sport, 72 Cabriolet.**
Row 4: **70 Matford Milord, 71 Graham-Paige, 73 Conduite Intérieure, 76 Chevrolet Coupe.**

13

Row 1: **Junior 77 Packard, 78 Autocar, 79 Tatra, 80 Alfa Romeo.**
Row 2: **81 Delahaye, 82 Ferrari, 83 Chevrolet, 84 Studebaker.**
Row 3: **88 Oldsmobile, 89 Nash, 90 Cadillac, 91 Renault Frégate.**
Row 4: **92 Pompiers (x2), 96 Grande Echelle (= men of #95), 113 Ford Comète.**

Row 1: **Junior 114 Packard, 115 Studebaker, 116 Ford Vedette, 195 Thunderbird.**
Row 2: **191 Mercury (x2), 192 Simca Versailles, 193 Peugeot 403.**
Row 3: **196 Ford Wagon & Ambulance, 197 Ford Wrecker & Pickup.**
Row 4: **207 Simca Chambord, 210 Rolls-Royce, #? Pickup & Animal Trailer.**

Row 1: **Junior 98 Lambretta, 99 Triporteur (front), 101 Terrot, Baby 130 Tractor & Trailer.**
Row 2: **122 Berline, 124 Limousine, 129 Coupé Sport, 131 Voiture Aerodynamique.**
Row 3: **133 Peugeot 203, 135b Peugeot 203, 136 Simca Aronde, 126 Limousine (x2).**
Row 4: **137 Nash, 138 Tanker, 140 Henry J (x2), 141 Van.**

Row 1: **Mosquito 151 Citroen 11 CV, 152 Maserati (x2), 153 Renault Frégate, 154 Hotchkiss, 155 Peugeot 203.**
Row 2: **156 Renault Pickup, 157 Ford Vedette, 159 Simca 9 (x2), 160 Tatra (x2), 161 Ford Comète.**
Row 3: **100 Jaguar D (x3), 101 Porsche Spyder, 102 Maserati 250F1.**
Row 4: **103 Ferrari TRC, 104 Vanwall, 105 Mercedes 190SL (x2), 106 Alfa Romeo Giulietta.**

17

Row 1: **107 Aston Martin DBR1 (x3), 113 Fiat-Abarth Record (x2).**
Row 2: **109 Renault Floride, 110 Simca Oceane (x2), 111 Aston Martin DB4, 112 DB-Panhard.**
Row 3: **108 Peugeot 403 (x2), 113b Ford Thunderbird, 115 Rolls-Royce.**
Row 4: **114 Citroen Ami-6, 116 Cooper Formula 2 (x4).**

Row 1: **119 Chausson Bus (x2), 120 Chausson Trolleybus.**
Row 2: **117 Porsche F2, 118 Lotus 18 (x2), 121 Lancia Flaminia (x2).**
Row 3: **122 Ferrari 156, 123 Ferrari 250GT, 124 Abarth 1000 (x3).**
Row 4: **125 Alfa Romeo 2600 (x4), 127 NSU Prinz IV.**

Row 1: **126 Mercedes 220SE, 131 BRM F1 (x3), 130 Aston Martin Vantage.**
Row 2: **130 Aston Martin Vantage (x4).**
Row 3: **128 Ford Thunderbird, 133 Fiat 2300S Ghia (x3).**
Row 4: **134 Porsche Le Mans (x2), 135 Lola F1 (x2), 129 Ferrari 2.5 liter.**

Row 1: **138 Harvey Aluminum Spl. (x2), 139 Maserati 3.5 (x2), 142 Alpine F3.**
Row 2: **141 Citroen Ami-6 Break, 143 Panhard 24BT (x3).**
Row 3: **146 Ford GT40, 147 Ford Mustang (x2), 147b Mustang Rally.**
Row 4: **148 Alfa Romeo GTZ (x2), 150 Oldsmobile Toronado (x2).**

Row 1: **151 Porsche Carrera 6 (x3), 152 Ferrari 330 P3, 153 Chapparal 2D.**
Row 2: **157 BMW 2000 CS, 157b BMW 2000 CS Rally (x2), 168 Alpine.**
Row 3: **161 Lamborghini Miura (x4), 169 Chapparal 2F.**
Row 4: **164 Simca 1000 (x3), 166 DeTomaso Mangusta (x2).**

Row 1: **165 Ferrari GTB4 (x3), 174 Porsche 908, 167b Ferrari F1.**
Row 2: **171 Opel GT (x3), 167b & 167 Ferrari F1.**
Row 3: **170 Ford Mark IV (x4).**
Row 4: **172 Alfa Carabo, 173 Matra F1, 175 Lola T70 (x2), 176 McLaren M8B.**

Row 1: **177 Ferrari 312P, 178 Matra 650, 179 Porsche 914-6, 181 Alpine (x2).**
Row 2: **180 Mercedes C-111 (x2), 184 Citroen SM (x2).**
Row 3: **182 Ferrari 512S (x2), 183 Alfa Romeo Junior (x2), 187 Alfa Romeo 33/3.**
Row 4: **186 (x2) & 186M Porsche 917, 188 Opel Manta (x2).**

Row 1: **185 Maserati Indy (x3), 189 Bertone Buggy (x2).**
Row 2: **190 Ford Capri, 192 Alpine A310 (x4).**
Row 3: **192b Alpine Police, 193 Citroen GS (x3), 194 Ferrari 312BB.**
Row 4: **195 Ligier JS3, 196 Renault 17TS (x4).**

Row 1: **197 & 197b Ferrari 512M, 198 Porsche 917, 199 March 707, 15 Lola T280.**
Row 2: **13 short & 14 long Matra 670, 10 Renault 5TL (x3).**
Row 3: **12 Peugeot 104 (x5).**
Row 4: **16 Ferrari Daytona (x3), 20 Alpine A441.**

Row 1: **17 Ford Mirage, 19 VW Golf (x2), 24 Porsche Carrera (x2).**
Row 2: **18 Porsche 917/10 (x3), 21 Matra Bagheera (x2).**
Row 3: **22 Renault 12 Break (x5).**
Row 4: **23a, 23b, 23c & 23d Peugeot 504 Break.**

Row 1: **25 BMW 3.0 CSL, 26 Ford Capri, 27 Lancia Stratos (x2), 36 Porsche 914-6.**
Row 2: **28 BMW 2002 (x2), 37 Renault 17TS, 34 Simca 1100 TI (x2).**
Row 3: **39 Simca 1308 GT (x4).**
Row 4: **29 Citroen CX 2200, 33 Fiat X1/9 (x3), 38 Gulf-Ford GR8.**

Row 1: **39 Simca 1308 GT (x4), 42 Renault 4 Van.**
Row 2: **42 Renault 4 Van (x5).**
Row 3: **40 Peugeot 604 (x3), 45 Ford Escort (x2).**
Row 4: **41 Alfa Romeo 33, 43 Renault 14 (x2), 44 Ferrari BB (x2).**

Row 1: **47 Mercedes 280E (x4).**
Row 2: **50 Peugeot 504 Rally, 47 Mercedes 280E (x3).**
Row 3: **49 Porsche 928 (x3), 53 Ford Fiesta (x2).**
Row 4: **52 Lancia Beta (x2), 54 Fiat-Abarth 131 (x2), 61 Ford Escort.**

Row 1: **56 Citroen 2CV6 (x2), 57 Alpine A442, 58 Renault 5 Gordini (x2).**
Row 2: **60 Simca 1308 Taxi (x2), 69 Alfasud Trofeo (x2), 70 Opel Kadett.**
Row 3: **63 Porsche 911 (x3), 68 Porsche 934 (x2).**
Row 4: **65 Citroen CX Break (x2), 66 Land Rover (x2).**

Row 1: **72 Citroen LN (x2), 75 BMW 3.0 CSL, 82 Alfetta GTV, 86 Porsche 936.**
Row 2: **73 & 73b Lancia Stratos, 76 Simca Horizon (x2), 87 Alpine A442B.**
Row 3: **81 Peugeot 104ZS, 89 BMW 530 (x3), 90 Peugeot 305.**
Row 4: **91 Renault 18 (x2), 96 Jaguar XJ12 (x2).**

Row 1: **200 M-20 Combat Car, 201 Unic Rocket Truck, 202 Patton Tank.**
Row 2: **206 Howitzer, 204 & 205 105 cc Cannon.**
Row 3: **207 PT76 Amphibian Tank, 208 SU100 Tank, 209 AMX 30 Tank.**
Row 4: **211 Berliet Tank Transporter (x2).**

Row 1: **218 PT76 Rocket Tank, 222 Tiger Tank, 219 M-41 Tank Destroyer.**
Row 2: **212-213 Auto-Union Jeep & Trailer, 203 Renault 4x4 Truck, 214 Berliet Aurochs, 221 Alfa Romeo Police.**
Row 3: **223 AMX AA Tank, 224 XM706 Amphibian, 224b XM706 Police, 225 BTR40 Rocket Carrier.**
Row 4: **227 AMX Ambulance, 227b AMX VTT, 226 Büssing Scout Car, 228 Jagdpanther.**

Row 1: **233 Renault R35, 230 AMX 13 Tank, 231 Sherman Tank, 234 Somua Tank.**
Row 2: **237 Panzer IV Tank, 232 M-10 Tank Destroyer, 240s & 240 Panhard AML.**
Row 3: **241 Hanomag Halftrack, 242 Dodge Truck, 235 Simca 4x4 & 105 Cannon.**
Row 4: **244 M-3 Halftrack, 243 Leopard Tank, 245 Kaiser 6x6 Truck.**

Row 1: **247 Berliet Alvis, 252 M7B1 Priest, 262 Richier Crane.**
Row 2: **249 AMX 13 Tank, 257 Saviem Tanker, 259 Citroen Ambulance.**
Row 3: **255 Berliet Foam Truck, 253 General Lee Tank, 256 Jeep & Trailer.**
Row 4: **250 AMX 13 Tank, 251 Saviem VAB, Saviem Gendarmerie (from #7020).**

Row 1: **300 Berliet TBO Petrolier, 303 Berliet Dump Truck.**
Row 2: **Unic Sahara Cement Truck, 303 Berliet Dump Truck.**
Row 3: **302 Willeme Horizon, 304 Bernard Refrigerator Truck.**
Row 4: **305 Berliet T12SR Low Loader.**

Row 1: **321 Saviem Auto Transporter & Trailer.**
Row 2: **316 Saviem 300SR with Crane, 306 Berliet Stradair Dumper.**
Row 3: **316 Saviem 300SR with Crane, 306 Berliet Stradair Dumper.**
Row 4: **317 Berliet Yoplait Semi, 307 Berliet Stradair.**

Row 1: **330 Citroen Publicity Van, 334 Richier Crane, 331 Mercedes Covered Truck.**
Row 2: **332 Saviem Cage Truck, 333 DAF Box Office.**
Row 3: **335 DAF Animal Truck, 337 DAF Caravan Truck.**
Row 4: **336 DAF Stake Truck, 338 Stake Trailer.**

Row 1: **318, 319 & 320 Saviem Elf, Esso & Shell Tankers.**
Row 2: **308 Willeme Elf Tanker, 350 Berliet Fire Truck, 351 Berliet Foam Truck.**
Row 3: **352 Berliet Ladder Truck, 353 Richier Crane (x2).**
Row 4: **354 Berliet Forest Fire Truck & Trailer, 355 & 355b Peugeot Bus, 359 Simca-Unic Snowplow.**

Row 1: **356 Volvo Dumper, 357 Unic Dumper, 364 Mercedes Bucket Truck.**
Row 2: **358 Mercedes Overhead Truck (x2), 364 Bercedes Bucket Truck.**
Row 3: **363 Magirus Fruehauf Semi, 361 Mercedes Ladder Truck.**
Row 4: **363 Magirus Fruehauf Semi, 367 Volvo Shovel Loader.**

Row 1: **362 Hotchkiss Fire Truck, 363b DAF Onatra Semi, 368 Citroen Fire Ambulance.**
Row 2: **369 DAF Shell Tanker, 371 Citroen Citroen Ambulance & Lifeboat.**
Row 3: **370 Saviem Renault Semi, 373 Mercedes Livestock Truck, 372 Peugeot Police Bus.**
Row 4: **366 Saviem SOS, Police & Fire Wreckers, 375 Berliet Fire Truck.**

Row 1: **374 Iveco Dump Truck (x2) flanking 378 Mercedes Excavator.**
Row 2: **376 Mercedes Bulk Carrier, 385 Saviem Horse Van.**
Row 3: **386 Mercedes Propane Tanker, 379 Mercedes Garbage Truck, 380 Peugeot Fire Ambulance (x2).**
Row 4: **388 Mercedes Stake Semi, 389 Stake Trailer.**

43

Row 1: **510 & 511 = 512 Renault Tractor & Tipping Trailer, 510 Renault Tractor, 515 Silage Trailer.**
Row 2: **510 & 513 = 514 Renault Tractor & Tank Trailer, 516 Sprayer Trailer, 613 Police Set.**
Row 3: **Truck made of parts from sets, 384 Mercedes Covered Truck, Junior 158 Caravan.**
Row 4: **391 Dodge Truck & Trailer, 621 Circus Caravan Set.**

Row 1: **600 Peugeot 504 (x2), 602 Renault 16.**
Row 2: **611 Opel Commodore (601) & Boat Trailer, 620 Wrecker Set.**
Row 3: **618 Racing Team Set, 619 Renault & Boat Trailer.**
Row 4: **615 Fiat X1/9 & Motorcycle, Solido-Marx Ferrari P3 & Porsche Carrera 6.**

Row 1: **1023 Renault 5 Turbo, 1031 BMW M1, 1032 Porsche 935 (x2).**
Row 2: **1034 Fire Land Rover, 1051 Porsche 924 (x2), 1055 Peugeot 504.**
Row 3: **1059 VW Scirocco, 1062 Matra Rancho (x2), 1094 Toyota Celica.**
Row 4: **1096 Jaguar XJ12 (x3), 1097 Porsche 934.**

Row 1: **1202A Fiat Ritmo, 1207A Ford Escort, Cougar 1301 Citroen 2CV6 (x2).**
Row 2: **Cougar 1302 Citroen Visa (x2), 1303 Fiat Ritmo (x2).**
Row 3: **Cougar 1304 BMW 530 (x3), 1313 Ford Fiesta.**
Row 4: **Cougar 1305 Alfetta GTV (x2), 1306 Peugeot 504 (x2).**

Row 1: **1302 Citroen Visa (x5).**
Row 2: **Citroen 2CV6 (x2), 1303 Fiat Ritmo (x3).**
Row 3: **1303 Fiat Ritmo Doctor, 1305 Alfetta GTV (x4).**
Row 4: **1304 BMW 530 (x2), 1307 Talbot Tagora (x2).**

Row 1: **1306 Peugeot 504 (x4).**
Row 2: **1308 Renault Fugeo (x2), 1309 Renault 14, 1312 Peugeot 505, 1313 Ford Fiesta.**
Row 3: **1310 Alfasud (x5).**
Row 4: **1315 Ford Escort (x3), 1316 Peugeot 104, 1317 Renault 5.**

Row 1: **1318 Renault 18 (x2), 1319 Talbot Horizon, 1320 Peugeot 305 (x2).**
Row 2: **1321 Renault 5 Turbo (x4), 1324 Porsche 924.**
Row 3: **1323 Porsche 934, 1327 Lancia Rally (x3).**
Row 4: **1322 Fire Jeep, 1325 Renault 4L Van (x3), 1329 BMW M1.**

Row 1: **1328 Audi Quattro (x2), 1330 Renault 4L Van, 1331 Jeep Rally.**
Row 2: **1332 Porsche 935, 1333 Alpine A442, 1334 Porsche 936, 1335 Peugeot 504.**
Row 3: **1336 Porsche 928 (x2), 1337 Mercedes 190 (x2).**
Row 4: **1338 Chevrolet Camaro, 1339 Renault 25 (x3).**

Row 1: **1340 Ford Sierra (x3), 1351 Peugeot 205, 1353 Renault 5 Maxi.**
Row 2: **1341 Nissan Prairie (x2), 1348 Porsche 944, 1359 Citroen 2CV6, 1350 Ford Escort RS.**
Row 3: **1349 Peugeot 205 GTI (x4), 1352 Mercedes 190.**
Row 4: **1354 Alfetta GTV, 1355 BMW M1, 1357 Renault Super 5, 1358 VW Golf, 1365 Citroen CX Ambulance.**

Row 1: **1501 Jaguar XJ12, 1502 Porsche 944 (x2), 1503 Alpine A310.**
Row 2: **1504 Renault 25, 1505 Porsche 928S (x2), 1506 Mercedes 190.**
Row 3: **1507 Chevrolet Camaro (x2), 1508 Peugeot 205 (x3).**
Row 4: **1509 Chevrolet Camaro, 1510 Mercedes 190, 1511 Rolls-Royce Corniche (x2).**

Row 1: **1512 Bentley, 1513 & 1514 Corvette, 1515 Ferrari 512BB.**
Row 2: **1516 Peugeot 605, 1517 & 1518 Mercedes SL, 1521 BMW 3.**
Row 3: **1519 & 1520 Renault Clio, 1522 Renault Espace (x2).**
Row 4: **1523 & 1524 Citroen ZX, 1525 Porsche 928, 1525 Renault Clio.**

Row 1: **1702 Porsche 935, 1703 Porsche 934, 1704 Fiat Abarth 131, 1705 Lancia Stratos.**
Row 2: **1708 BMW 530, 1712 Porsche 934, 1801 Maserati Indy, 1806 Jaguar XJ12.**
Row 3: **1802 Ferrari BB, 1803 & 1804 Alpine A110, 5019 VW Golf (x2).**
Row 4: **1805 Opel GT, 5041 Alfa Romeo 33, 5050 Peugeot 504.**

Row 1: **5058 Renault 5, 5068 Porsche 934, 5069 Alfasud TI, 5081 Peugeot 104ZS (x2).**
Row 2: **5089 BMW 530, 5094 Toyota Celica (x2), unknown Porsche 934.**
Row 3. **2401 & 2402 DB Panhard, 2405 & 2406 Alpine.**
Row 4: **2403 & 2404 Ferrari GTO, 2407 & 2408 Matra 670.**

Row 1: **2409 & 2410 Ferrari GTB4, 2411 & 2412 Porsche 935.**
Row 2: **2413 & 2414 Ferrari TR, 2415 & 2416 Ford Mark IV.**
Row 3: **2417 & 2418 Ligier JS3, 2419 & 2420 Alfa Romeo 33/3.**
Row 4: **2421 & 2422 Porsche 917, 2423 & 2424 Lola T280.**

Row 1: **2001 Saviem Fire Truck, 2002 Citroen C35 Van, 2004 Matra Rancho, 2005 Land Rover.**
Row 2: **2006 Saviem Wrecker, 2007 Dodge Truck, 2105 Dodge Fire Truck, 2110 Volvo Roller.**
Row 3: **2108 Simca-Unic Snowplow, 2113 Simca-Unic Forest Fire Truck, 2115 Peugeot J7 Bus (x2).**
Row 4: **2116 Citroen Ambulance (x2), 2118 Citroen Fire Van, 2119 Matra Rancho.**

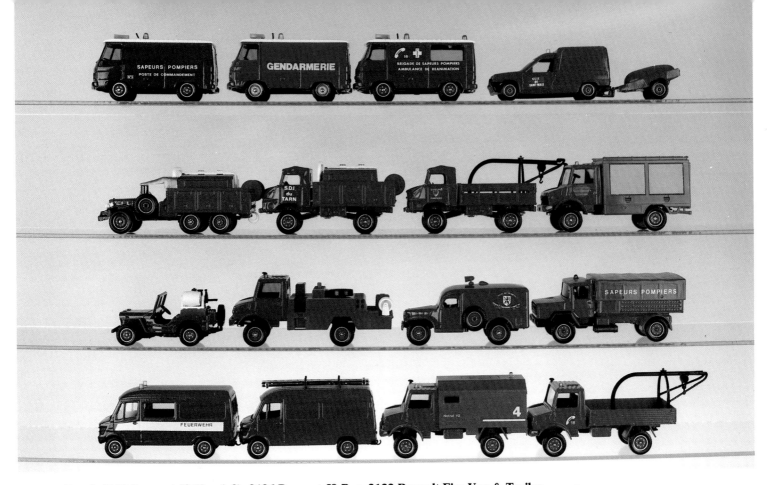

Row 1: 2120 Peugeot J9 Van (x2), 2126 Peugeot J9 Bus, 2122 Renault Fire Van & Trailer.
Row 2: 2123 Dodge Fire Tanker, 2121 Simca-Unic Fire Truck, 2124 Simca-Unic Wrecker, 2125 Fire Unimog.
Row 3: 2117 Fire Jeep, 2127 Forest Fire Unimog, 2128 Dodge Fire Van, 2129 Iveco Fire Truck.
Row 4: 2131 Mercedes Rescue Van, 2132 Mercedes Fire Van, 2133 Unimog Fire Ambulance, 2134 Unimog Fire Wrecker.

Row 1: **3001 Land Rover & Cage Trailer, 3002 Saviem Wrecker & Renault 5.**
Row 2: **3003 Matra Rancho & Horse Trailer, 3004 Dodge Truck & Trailer, 3007 Iveco Esso Truck.**
Row 3: **3005 Simca-Unic Snowplow & Trailer, 3000/3100 Citroen Van & Lifeboat.**
Row 4: **3102 Richier Crane, 3107 Berliet Foam Truck, 3110 GMC Fire Crane.**

Row 1: **3108 Mercedes Overhead Truck, 3111 Mercedes Ladder Truck, 3112 Berliet Overhead Truck.**
Row 2: **3106 Mack Fire Truck (x3).**
Row 3: **3113 GCM Open Truck, 3114 Mercedes Fire Truck, 3115 GMC Smoke Ejector.**
Row 4: **3116 GMC Smoke Ejector, 3117 GMC Fire Crane, 3118 Iveco Fire Truck.**

Row 1: **3119 & 3120 Sides-Mack Airport Fire Trucks, 3306 Skip Dumper.**
Row 2: **3121 GMC Covered Truck, 3122 Mercedes Van & Lifeboat, 3304 Fire Jeep & Trailer.**
Row 3: **3303 Saviem Fire Truck, 3307 Iveco Esso Truck, 3500 DAF Shell Tanker.**
Row 4: **3500 DAF Texaco & Elf-Antar Tankers.**

Row 1: **3501 DAF Danzas Semi, 3502 Saviem Saunders Semi.**
Row 2: **3501 DAF Kvas Semi, 3502 Saviem Renault Semi.**
Row 3: **3501 DAF Opal Semi, 3503 Saviem Pipe Truck.**
Row 4: **3504 Iveco Ferreri Semi, 3507 DAF Truck & Trailer.**

Row 1: **3505 Mercedes Bulk Carrier, 3510 Mercedes Container Semi.**
Row 2: **3506 Iveco Milk Tanker, 3508 Mack Mayflower Semi.**
Row 3: **3506 Iveco Tecni Plast Tanker, 3508 Mack Husqvarna Semi.**
Row 4: **3509 Renault Fire Tanker, 3511 Mack Fire Truck.**

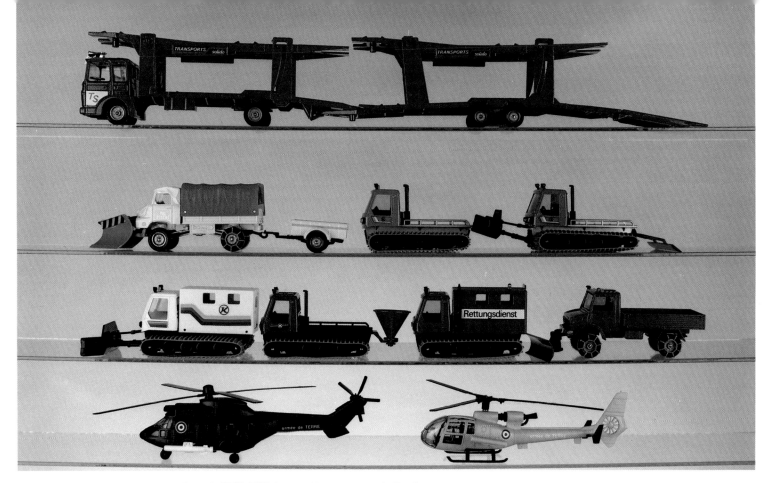

Row 1: **3700/7006 Auto Transporter & Trailer.**
Row 2: **3600 Simca-Unic Snowplow & Trailer, 3601 Kässbohrer Pistenbully (x2).**
Row 3: **3602, 3603 & 3607 Kässbohrer, 3606 Unimog with Snowplow.**
Row 4: **3828 Cougar & 3829 Gazelle Armée de Terre Helicopters.**

Row 1: **3810 Gazelle Europ'Assistance, 3811 Gazelle Gendarmerie, 3812 Gazelle Carabinieri.**
Row 2: **3813 Gazelle Policia, 3814 Alouette Security, 3815 Gazelle Armée de Terre.**
Row 3: **3821 Alouette, 3822 Gazelle, 3823 Alouette Gendarmerie.**
Row 4: **3824 Puma Armée de Terre, 3825 Puma Aerospatiale.**

Row 1: **154/4154 Fiat 525N (x3), 4157 Fiat 525N.**
Row 2: **137/4137 Mercedes SS (x4).**
Row 3: **4001 Mercedes SSKL (x2), 132/4132 Mercedes SS (x2).**
Row 4: **4004 Mercedes SSKL (x2), 140/4140 Panhard 35 CV (x2).**

Row 1: **145/4145 Hispano-Suiza (x3), 144/4144 Voisin.**
Row 2: **62/1162/4062 Hispano-Suiza (x3), 136/4136 Bugatti.**
Row 3: **4088 Bugatti, 4109 Bugatti (x2), 4036 Bugatti.**
Row 4: **88/4088 Bugatti (x4).**

Row 1: **146/4146 Duesenberg (x4).**
Row 2: **146/4146 Duesenberg (x4).**
Row 3: **35/4035 Duesenberg (x3), 4086 Mercedes 540K.**
Row 4: **67/4067 Mercedes 540K (x2), 4086 Mercedes 540K (x2).**

Row 1: **4003 Talbot T23 (x4).**
Row 2: **59/1159/4059 Renault 40CV (x4).**
Row 3: **149/4149 Renault 40CV (x3), 4159 Ford V8.**
Row 4: **97/4097 Renault Reinastella (x4).**

Row 1: **48/1148/4048 Delahaye 135M (x5).**
Row 2: **4048 (x2) & 78/4078 (x2) Delahaye 135M.**
Row 3: **51/1151/4051 Delage D8-120 (x3), 78 Delahaye 135M.**
Row 4: **31/4031 Delage D8-120 (x4).**

72

Row 1: **77/4077 Rolls-Royce Drophead (x4).**
Row 2: **4077, 46/4046 (x2) & 71 Rolls-Royce.**
Row 3: **71/4071 Rolls-Royce Town Car (x4).**
Row 4: **71/4071 Rolls-Royce Town Car (x4).**

Row 1: **85/4085 Cadillac V16 (x4).**
Row 2: **4085 Cadillac V16 (x2), Cadillac Hewlett-Packard & Expo 92 vans.**
Row 3: **Cadillac Castlemaine Perkins, 4070 Fire Chief, 4075 Sellers Fire and Zellers vans.**
Row 4: **Cadillac Queru, 4060 Cadbury, 4061 Banania & 4065 Waterman vans.**

Row 1: **4002 Jaguar SS 100 (x2), 4099 Packard 1937 (x2).**
Row 2: **4037 (x2) & 4047 Packard 1937.**
Row 3: **4047 Packard 1937 (x3).**
Row 4: **Cadillac 4042 Denver, 4043 & 4057 Police, 4038 Manhattan Fire.**

Row 1: **55/4055 Cord L29 (x4).**
Row 2: **80/4080 Cord L29 (x4).**
Row 3: **80/4080 Cord L29 (x3), 4431 Ford V8 Van.**
Row 4: **Budweiser Set: 9181 Dodge Flatbed, 9180 Cadillac Van, 9183 Citroen Van, 9182 Dodge Pickup.**

Row 1: **4115 & 4526 Citroen 15CV, Citroen 7th & 8th ICCCR promos.**
Row 2: **4033 Citroen Fire Chief (x2), 32B & 4040 Citroen FFI.**
Row 3: **32/4032 Citroen 15CV (x3), 4041 Citroen Taxi.**
Row 4: **4032 50th Anniversary Citroen (x3), 4519 Citroen 15CV.**

Row 1: **4402 AEC Doubledeck Bus (x2), flanking 4403 Citroen C4 Fire Truck.**
Row 2: **4401 & 4406 Renault Paris & Lyon Buses.**
Row 3: **4405 Citroen Hotel Bus (x2), 4407 Michelin Van, 4409 Samaritane Van.**
Row 4: **4408 Citroen Coal Truck (x2), 4410 Citroen Wrecker (x2).**

Row 1: **4404 AEC Doubledeck Bus (x2), flanking 4412 Dodge Flower Shop.**
Row 2: **4413 Dodge Pickup, 4414 Dodge Coal Truck, 4420 Dodge Fire Pickup, 4421 Dodge Sun Club Truck.**
Row 3: **4415 Dodge Fire Tanker, 4418 Dodge Milk Tanker, 4419 & 4426 Dodge Texaco Tankers.**
Row 4: **4423 Dodge Open Truck, 4424 Dodge Wrecker (x2), 4428 Dodge Fire Wrecker.**

Row 1: **4417 AEC Open Top Bus, 4422 Citroen Fire Tanker, 4402 AEC Doubledeck Bus.**
Row 2: **4411 Citroen Ambulance, 4416 Citroen BP Van, 4429 Citroen Kodak Van, 9189 Citroen Club de l'Auto Van.**
Row 3: **4425 Dodge Fire Truck, 4427 Dodge Pepsi-Cola, 4430 Dodge Sunlight Soap, Dodge Marines Ambulance.**
Row 4: **Citroen Tetleys Ales, Galeries Lafayette, Solido & British Meat vans.**

Row 1: **CS-2 Duesenberg J, CS-1 Spark Plug, Cadillac Boston Fire Car.**
Row 2: **Four Citroen Limited Edition fire trucks.**
Row 3: **Four Dodge Limited Edition fire trucks.**
Row 4: **Citroen 15CV, Citroen Wrecker, Chevrolet & Chrysler Fire Chief cars.**

Row 1: **9601 Cadillac, 9701 AEC Doubledeck Bus, 9610 Cadillac Eldorado Coca-Cola.**
Row 2: **9602 Chevrolet, 9603 Dodge, 9604 Chrysler, 9605 Dodge Coca-Cola.**
Row 3: **9606 Citroen, 9607 Chevrolet, 9608 Chrysler, 9609 Dodge Coca-Cola.**
Row 4: **Ford Thunderbird & Chevrolet Coca-Cola, Citroen Expo 92 & AKAI vans.**

Row 1: **4500 Cadillac Eldorado Convertible (x3), 4502 Mercedes 300SL.**
Row 2: **4501 Cadillac Eldorado Hardtop (x3), 4502 Mercedes 300SL.**
Row 3: **4500 Cadillac Eldorado, 4505 Ford Thunderbird Hardtop (x3).**
Row 4: **4504 Ford Thunderbird Convertible (x4).**

Row 1: **4503 Mercedes 300SL (x2), 4505 Ford Thunderbird Hardtop (x2).**
Row 2: **4506 Ferrari 250GT (x2), 4505 Ford Thunderbird, 4508 Chevrolet.**
Row 3: **4507 Ferrari 250GT (x2) 4509 Chevrolet Taxi, 4510 Chevrolet Police.**
Row 4: **4508 Chevrolet 1950 (x4).**

Row 1: **4511 Buick Convertible (x2), 4513 Chrysler Windsor (x2).**
Row 2: **4512 Buick Convertible (x3), 4523 Buick Hardtop.**
Row 3: **4514 Chrysler Taxi, 4518 Chevrolet Fire Chief, 4520 Cadillac Hardtop (x2).**
Row 4: **4515 Facel Vega Hardtop (x2), 4516 Facel Vega Cabriolet (x3).**

Row 1: **4521 & 4522 Studebaker Silver Hawk, 4517 Ford Thunderbird.**
Row 2: **4524 Tucker 1948 (x4).**
Row 3: **4524 Tucker 1948 (x3), 4529 Chevrolet Taxi.**
Row 4: **4525 Chrysler Fire Chief, Chevrolet Sheriff and RCMP, 4530 Chrysler Police.**

Row 1: **7006 Auto Transporter & Trailer.**
Row 2: **7013 Berliet Low Loader.**
Row 3: D-Day **603 M-10 Tank Destroyer, 604 Tiger Tank, 605 Sherman Tank, 607 M-20 Combat Car.**
Row 4: D-Day **606 Panther Tank, 608 M3 Halftrack, 609 Dodge Ambulance, 610 Büssing Scout Car.**

Row 1: **Overlord 1 Jeep & Trailer, 2 GMC LeRoi, 8 M-20 Combat Car, 7 Dodge Truck.**
Row 2: **Overlord 3 & 4 GMC Trucks, 5 Dodge Ambulance, 6 Dodge Signal Van.**
Row 3: **Overlord 9 Packard Staff Car, 10 M-3 Halftrack, 11 Hanomag Halftrack, 12 Büssing Scout Car.**
Row 4: **6001 GMC Compressor, 6002 GMC Wrecker, 6003 Cadillac Staff Car, 6004 Dodge Signal Van.**

Row 1: **6005 Kaiser Truck, 6006 Packard Staff Car, 6020 Citroen Ambulance, 6021 Citroen Truck.**
Row 2: **6022 Simca-Unic Truck, 6023 Renault Truck, 6028 M-20 Combat Car, 6025 Panhard AML, 6030 Renault 4x4.**
Row 3: **6026 Commando Car, 6007 & 6027 VAB, 6027 VAB UN.**
Row 4: **6024 Dodge Ambulance, 6029 Kaiser Truck, 6032 GMC Truck, 6031 BTR Rocket Launcher.**

Row 1: **6033 Chevrolet Staff Car, 6034 Jeep & Trailer, 6035 Citroen FFI, 6037 DKW & Trailer.**
Row 2: **6036 GMC Truck, 6038 Unimog Red Cross, 6039 Land Rover & Trailer, 6040 Dodge Pickup.**
Row 3: **6041 Jeep & Boat, 6042 Chrysler Staff Car, 6043 Dodge Ambulance, 6044 GMC Compressor.**
Row 4: **6045 GMC Wrecker, 6046 Unimog Ambulance, 6047 GMC Truck, 6048 Jeep & Trailer.**

Row 1: **6049 Machine-Gun Jeep, 6051 Büssing Scout Car, 6052 M-3 Halftrack, 6055 Leopard Tank.**
Row 2: **6053 Sherman Tank, 6054 AMX 10, 6056 VAB 6x6, 6057 AMX 13 VTT.**
Row 3: **6058 & 6059 AMX 13 Tanks, 6060 AMX 30 Tank, 6061 Hanomag Halftrack.**
Row 4: **6062 AMX Ambulance, 6063 Tiger Tank, 6064 Jagdpanther (x2).**

Row 1: **6065 Patton Tank, 6066 AMX Lance-Rocket, 6067 General Lee Tank, 6072 Renault R35 Tank.**
Row 2: **6068 M-10 Tank Destroyer, 6069 Recovery Halftrack, 6070 Kaiser Crane Truck.**
Row 3: **6071 General Grant Tank, 6073 Panzer IV Tank, 6074 Somua Tank, 6075 PT76 Tank.**
Row 4: **6076 AMX 10, 6077 Sherman Tank with Plow, 6078 Sherman Tank, 6079 AMX 30 Tank.**

Row 1: **163 Mystère IV, 164 Fouga Magister, 165 Skyray, 167 Leduc 021, 168 Super Sabre, 170 MIG-15.**
Row 2: **172 Thunderjet, 173 Baroudeur, 177 Super-Cigale, 187 Trident, 179 Fairey Delta, 180 Convair.**
Row 3: **182 Morane-Saulnier, 183 Aquilon, 184 Super Mustère, 185 Etendard IV.**
Row 4: **181B Vertol Helicopter, 187 Tupolev TU-104.**

7028 Gendarmerie set

Tour Auto Set

7015 Mountain Set

7035 Aeroports De Paris Set

7002 Fire Vehicle Set

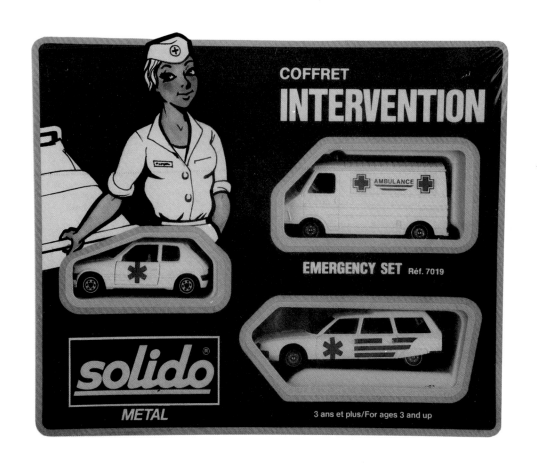

7019 Intervention (Emergency) Set

MAJOR SERIES

The production of the Bougie Gergovia spark plug on wheels, which has (belatedly, we suspect) been given catalog number 0, was followed in 1932 by the introduction of the Major Series of more-or-less 1/35 scale models. They were composed of a chassis, a body and hood which bolted to it, other metal parts such as seats which also bolted on, and four cast metal wheels. Only after the first thirteen models (#1-13) were produced did Solido decide to equip the models with clockwork motors, which also bolted to the chassis. To wind up the motors, a key was used--which necessitated a round keyhole through one side of the hood.

Speaking of the hood, there were several patterns available. One design had an absolutely vertical face, with a grille of vertical lines in a rectangle. Several other types were raked; some had slanted front faces with vertical lines, at least one having a squared bottom and one with a point at the base. Others were raked and divided into two halves by a central line, leaving two pancls in a wide V, pointed at the bottom. There was a two-piece hood composed of right and left castings, each of which supplied half of the grille, a rounded shield type, also with vertical lines. Finally, there was a very different rounded type, with the flat top of the hood extending out in a semicircle over several horizontal ridges that curved around the front of the hood. The hoods had a variety of louver designs on the sides. Some of these types may have been used only for the Major, some only for the Junior series, and some in both series. More information can be found in Azema Volume I.

The Major Series models share numerous body types with the Juniors, which makes them somewhat easier to describe, but also leads to confusion when one sees pictures of them. Unless one knows the size of the model, one often cannot tell a Major (which is about six inches long) from a somewhat smaller Junior. Nor can one predict which of the hood-grille types might be found with a given body, as most are interchangeable. In fact, some of them were sold in sets of parts that could be assembled as desired. So we shall describe the body type and leave the rest to chance.

Finally, the names by which these models have been known differ too. Here we begin with the first French names I found in print, which were already on my computer before I obtained a copy of Azema Volume I and found different names in use. At this point I am not sure which names, if any, are official.

0 BOUGIE GERGOVIA 1930

Spark plug on wheels, cast aluminum body and wheels, not a model of a vehicle as such. Colors: white, red, and dark blue. The yellow model with Club Solido lettering is a modern reissue.

1 BERLINE 4 PORTES (Conduite intérieure) 1931

Four-door sedan with boxy cab, four big doors, large rear trunk with slanted spare wheel. Colors: chrome, cream, red, yellow, light blue, dark blue, light gray.

2 TORPEDO 4 PORTES (Torpédo) 1931

Four-door convertible with windshield frame, front and rear seats, and large rear trunk with vertical spare wheel. Colors: chrome, red, yellow, green, olive green, light gray.

3 COUPÉ DE VILLE (Coupé) 1931

Town car with open front and enclosed rear seats, large rear trunk with vertical spare wheel. Colors: chrome, white, metallic red, dark red, light blue, silver blue, dark blue, light gray.

4 SPEEDSTER DE COURSE (Pointe de course) 1931

Sports-racing car with open cab, long pointed tail and cast-in racing number 1 behind door. Colors: chrome, red, yellow, light green, dark green.

5 PICK UP (Camion) 1931

Pickup truck with windshield frame, open cab and large open rear with wheel wells. Colors: chrome, red, yellow, light green, green, dark green, light blue, light gray.

6 AUTOCAR OUVERT (Autocar) 1931

Open bus with three seats and folded top at rear. Colors: chrome, red, yellow, light green, blue.

7 FOURGON "SOLIDO" (Fourgon) 1932

Panel truck with cast-in "Solido" logo; one type has a spare wheel attached to the left side just behind the driver's door.
Colors: chrome, dark red, yellow, green, light blue, tan.

8 CABRIOLET (Roadster) 1932

Convertible two-seater with windshield frame, folded top and rounded rear trunk with vertical spare wheel. Colors: chrome, white, red, silver red, red gold, dark red, yellow.

9 FAUX CABRIOLET (Cabriolet) 1932

"False convertible" two-door with irons to make it look like a convertible, plus slanted trunk +/- vertical spare wheel. Colors: chrome, metallic red, yellow, light green, metallic light green, light blue, blue, metallic dark blue.

10 ROADSTER (Torpédo Grand Sport) 1933

Sports roadster with open cab, long sloping tail squared at rear, and cast-in racing number 7. Colors: chrome, red, dark red, yellow, light green, metallic blue, dark blue.

11 COUPÉ SPORT (Berline de course aérodynamique) 1933

Hardtop coupe with window opening extending aft of door, long tail with spare wheel on its surface. Colors: chrome, red, metallic red, metallic dark red, yellow, light green, light blue.

12 COUPÉ 2+2 (Coupé sport aérodynamique) 1935

Two-door coupe with boxy cab, very little room for "+2", long sloping tail squared at rear, no spare. Colors: chrome, red, brick red, metallic red, maroon, light green, metallic light green, light blue.

13 LIMOUSINE (Limousine aérodynamique) 1935

Four-door fastback with side windows only in doors, rear window in sloping tail, no spare. Colors: chrome, red, metallic red, light green, metallic green, dark green, light blue, dark blue.

13B LIMOUSINE 1936
Four-door sedan with three windows on each side, keyhole in rear door, skirted rear fenders (special chassis), cast-in spare. Usually seen with pointed overhanging hood. Any colors of #13 may be possible.

14 BERLINE MONOCOQUE (Coupé monocoque) 1936
A two- or four-door sedan of some kind, with keyhole on the left. Colors: chrome, cream, red, yellow, tan, green and tan camouflage.

15 GRAHAM-PAIGE LIMOUSINE (Limousine monocoque) 1937
Sedan, first type with vertical-bar grille, second type with overhanging nose similar to Junior #71. Colors: chrome, metallic dark red, maroon, yellow, light green, light blue, green and tan camouflage.

16 AUTOCAR AVEC TOIT OUVRANT (Autobus français) 1937
An open-top bus. Colors include chrome with red upper deck.

17 AUTOBUS À IMPERIALE (Autobus anglais) 1937
A doubledeck bus. Colors include chrome, chrome/red.

24 VOITURE AMPHIBIE COUPÉ 2+2 1935
This is not listed by Azema, but the name sounds so fascinating that I hardly dare to translate it. An amphibian 2+2 coupe?

37 CAMION DE CAMPAGNE 1936
A typical field gun, a cannon with spoked wheels.

38 CANON LOURD-155 MM 1936
This appears to be another cannon.

39 CANON DE FORTERESSE 1936
This cannon has a shorter, thicker barrel.

40 CANON ANTI-AERIEN 1936
An anti-aircraft gun.

41 CANON OBUSIER 1936
A short- and wide-barreled howitzer.

42 CANON ANTICHAR 1936
An antitank gun with a long, very thin barrel.

43 TRACTEUR 1938
This actually seems to be the powerplant for the following toys, which run on tracks rather than tires.

43A CHAR 1938
An army tank with turret (one gun or two), powered by the #43 motor. Colors: dark green, dark green and dark brown camouflage, or burnished metal. The following bodies (pardon my English) can be attached to the motor in place of the tank body:

43B ANTI-AIRCRAFT GUN TRUCK 1938

43C CRANE TRUCK 1938

43D DUMP TRUCK 1938

43E FIRE LADDER TRUCK 1938

43F PLOW, ROLLER AND HARROW 1938
These, it appears, can be pulled by a tractor using the same motor as the tank.

JUNIOR SERIES

This 1/40 scale series began in 1933 as a line of scaled-down counterparts of the Majors, but it lasted a lot longer--into the Fifties, in fact--and took on new forms. The prewar models were formed as were the Majors: of a chassis 100 to 110 mm long with four cast metal wheels, cast body, hood-grille and seats, and no motor; the few 1939 issues with motors (from #70 on) and the postwar models had a keyhole on the right or left side. In postwar days many models were available both with and without a motor, the two versions being sold under the names of two different French cities or districts. The postwar models also had metal hubs with rubber tires.

Many of the prewar Juniors had the same more or less generic body types and separate hood-grille castings as the Majors, but new types were also introduced, and the postwar models were generally "inspired by" (as some catalogs put it) real cars and can be recognized as such. Most postwar models followed the earlier tradition in having the body attached to the chassis by two bolts, but the postwar bodies were single castings, including the hood and grille. The very last type featured two-piece chassis completely enclosed in the bodywork--and a couple of these last bodies survived to join the 100 series that was introduced in 1957. It might be added that they deserved to, as the quality of these models improved visibly through the forties and fifties. So did their eye appeal in another sense; whereas many prewar Major and Junior models were chromed, the postwar Juniors were painted in bright yet fairly realistic colors. Some of them included removable plastic tops including windows, allowing a hardtop to be turned into a convertible just by lifting off the roof--the same idea seen later in the Dinky Toys E-Type Jaguar.

Since the last postwar Juniors were issued after the entire Baby Series, their numbers are not consecutive. The Juniors are:

56 COUPÉ DE VILLE 1933
Town car, similar to #3 but without the rear spare. Colors: chrome, red, dark green, blue, dark blue.

57 COUPÉ 2+2 (Conduite intérieure) 1933
Coupe, very much like #12; but #57 is also listed as:

57 BERLINE 4 PORTES year?
Four-door sedan, vaguely like #1 but shorter, with a more rounded roof and no spare wheel. Colors; chrome, red, orange, light green, dark green.

58 CABRIOLET (Torpédo Grand Sport) 1933

Roadster, rather like #8 but shorter and with no spare wheel. Colors: chrome, red, light green, light blue.

59 FOURGON "SOLIDO" (Fourgon) 1933

Panel van much like #7, with cast-in "Solido" logo, square rear, no spare wheel. Colors: yellow, green, dark green, light blue.

60 PICK UP (Camion) 1933

Pickup truck similar to #5. but without flared-out rear sides. Colors: chrome, red, yellow, light green, dark green, light blue, dark blue.

61 CAMION-CITERNE 1933

Tank truck unlike any Major, with three-section tank casting. Colors: chrome, red, yellow, olive green, dark blue.

62 AUTOCAR OUVERT (Autocar) 1933

Open bus very similar to #6, with three seats. Colors: chrome, red, yellow, light green, dark green, light blue, dark blue.

63 SPEEDSTER (Pointe de course) 1933

Sports-racing car somewhat like #4, with pointed tail, cast-in number 1, and no folded top behind the cockpit. Colors: chrome, red, yellow, light green, light blue, light gray.

64 ROADSTER 1933

Sports roadster like #10, with folded top. Colors: chrome, red, dark red, yellow, light gray.

65 COUPÉ SPORT (Berline Grand Sport) 1933

Sports coupe somewhat like #11, but with side windows only in the doors, not extending beyond them. Colors: chrome, cream, red, yellow, light green, metallic light green, light blue, blue.

66 AUTOCAR FERMÉ (Autocar) 1934

Singledeck bus, a new design with four windows on each side and cast-in "Grands Voyages" logo. Colors: chrome, red, yellow, light green, light blue.

67 BERLINE SPORT 1934

Fastback sedan, its two doors on each side having angled rather than vertical upright lines, and small side windows behind the rear doors. Some pictures indicate a raised trunk area, but mine is perfectly flush with the body lines. Colors: chrome, dark red, yellow, olive green, light blue, light gray.

68 SIMPLEX POMPIERS AVEC ÉCHELLE (Voiture de pompiers) 1939

Fire ladder truck, a new design with a raising ladder over a detailed rear body. Colors: red, dark red.

69 SIMPLEX BENNE À ORDURES 1939

Another closed-cab truck, with a long open rear ending in sides tapering upward. Colors: green, dark green, gray, metallic dark gray.

70 MATFORD CABRIOLET MILORD (Cabriolet décapotible) 1939

A sporty town car with open front seat; the body casting includes the hood, grille and windshield frame, while a second casting provides a cover for the back seat. Left keyhole. Touraine without, Normandie with motor. Colors: red, maroon, yellow, light green, light blue, blue, metallic blue, dark blue, gray.

71 GRAHAM-PAIGE COUPÉ (Coupé) 1939

A sleek coupe with single body casting, pointed and overhanging nose, horizontal grille lines and raised trunk. Left keyhole. Lorraine without, Ile-de-France with motor. Colors: red, metallic dark red, light green, olive green, dark green, light blue-green, dark blue.

72 BMW BARQUETTE (Cabriolet Sport) 1939

A long open sports car, the rear fenders of its body casting notched so the bumper ends can stick out of the overhanging tail, with a recognizable BMW grille and a solid cast-in windshield. Left keyhole. Dauphiné without, Aquitaine with motor. Colors: Gold, silver, white, red, yellow, yellow-orange, light green, light blue, blue, metallic blue, black.

73 BERLINE 4 PORTES (Conduite intérieure) 1939

If I've identified this correctly, it's an American-style four-door sedan with a single body casting featuring a pointed nose, a few horizontal grille lines, a central crest along the hood, and a left keyhole. Roussillon without, Gascogne with motor. Colors: cream, red, dark red, metallic dark red, yellow, olive green, dark olive green, dark green, sea blue, tan.

74 CAMION-PLATEAU AVEC GRUE 1939

A cab-chassis casting plus a flat rear body and a swiveling cast crane boom at the rear. Picardie without, Berry with motor. Colors: silver, red, dark red, red-orange, yellow, light green, light gray, silver and red, silver and brown-black.

75 SEMI-REMORQUE PLATEAU 1939

The cab-chassis from above, pulling a flat trailer with a crane at the front, and sometimes with stakes. Béarn without, Artois with motor. Same colors as #74.

76 COUPÉ 2 PORTES 1945

Back to single body castings for a more-or-less Chevrolet coupe with two big doors, a rounded cab rear with windows in Studebaker style, a hood ornament of sorts and a dull grille of two rectangles of horizontal lines. Left and right keyholes. Lorraine without, Ile de France with motor. Colors: red, yellow, pale green, pale gray.

77 LIMOUSINE (Conduite intérieure 4 portes) 1945

A four-door Packardish sedan with a Ford Vedette nose, Buick ratholes, Studebaker rear windows, a hood emblem with wings and another grille of horizontal lines in rectangles. Two keyholes.
Roussillon without, Gascogne with motor. Colors: White, red, dark red, yellow, green, dark blue.

78 AUTOCAR (Autocar de voyage) 1946

A streamlined singledeck bus with raised longitudinal roof section, two doors on each side, and one blocked-out window on each side where the motor goes. Two keyholes. Provence without, Auvergne with motor. Colors: red, brick red, dark red, yellow, light green, blue.

79 TATRA VOITURE À AILERON (Berline de compétition) 1946

A streamlined, futuristic two-door fastback with windows tapering away to nothing and a tailfin. The doors open directly on the motor, with the seat behind it. Two keyholes. Champagne without, Alsace with motor. Colors: white, red, red-orange, light yellow, golden yellow, light green, dark blue.

80 ALFA ROMEO CABRIOLET 1948

A pleasant open four-seater with recognizable grille, plain windshield frame and folded top, plus two keyholes. Hendaye without, Cannes with motor. Colors: red, yellow, olive green, metallic blue.

81 DELAHAYE CABRIOLET 1948

A nice two-seat convertible with distinctive grille, split windshield frame, big folded top, detailed trunk lid and two keyholes. Antibes without, Monaco with motor. Colors: white, cream, yellow, light green, light blue, metallic light blue, light gray.

82 FERRARI BARQUETTE 1948

A sleek open two-seater with rounded vertical-line grille, air scoop, two cast-in bucket seats, and plastic windshield, plus separate headlights. Two keyholes. Bourgogne without, Anjou with motor. Colors: chrome, white, red, dark red, yellow, light green, light blue.

83 CHEVROLET BERLINE 2 PORTES 1952

A two-door fastback that looks misshapen, as its tail is too low and its rear fenders too high. The motor pushes the seats too far back, which doesn't help either. Potato-chipper grille up front, plus separate headlights--this and the previous model are the only ones to have them; it was so much simpler to cast them in. Two keyholes. La Baule without, Houlgate with motor. Colors: cream, red, red-orange, maroon, yellow, light green, dark blue, purple, light gray, black.

84 STUDEBAKER COUPÉ (Coupé Grand Tourisme) 1952

A bit awkward but still a relief after the Chevy, with propeller-like nose, rounded rear and two keyholes. Bretagne without, Ardennes with motor. Colors: white, chrome, red, maroon, dark red, light green, dark blue, black.

85 BEAUCE TRACTEUR 1948-1963

A farm tractor with driver figure, it came in three successive varieties with slight structural differences. Colors: red, orange, yellow, light green, army green.

86 REMORQUE À FOIN 1948-1963

A group of farm machines for the tractor to pull: hay wagon, open farm wagon, roller, seeder, mower, harrow, etc. The assortment seems to have changed over the years.

86-A CHARRUE (Tracteur Flandres) 1948-1963

A similar tractor to #85, but running on tracks. Apparently the same colors.

86-B FAUCHEUSE 1948

A mower.

86-C BRISE-MOTTES TRIPLE 1948

A three-section harrow.

87 FLANDRE TRACTEUR À CHENILLES & ROULEAU 1948

A bulldozer-type tractor with a roller?

88 OLDSMOBILE BERLINE 1952

A four-door sedan with a crude grille and sufficient body line detail. Two keyholes. Brest without, Cherbourg with motor. Colors: dark red, metallic dark red, maroon, yellow, light green, dark blue, black, gray with gray-blue top.

89 NASH LIMOUSINE 1954

A very recognizable Nash with oval grille of vertical lines, fastback body, and two keyholes, but no motor. Nancy without motor. Colors: cream, red, maroon, light blue, metallic blue.

90 CADILLAC LIMOUSINE 1954

A solid four-door sedan with detailed grille, hood ornament and fenders, though its motor is too conspicuous and seat too narrow for the wide body. Two keyholes. Vichy without motor. Colors: metallic rose, maroon, pale green, dark green, dark blue, black.

91 RENAULT FRÉGATE 1954

A smooth four-door sedan with characteristic horizontal-line grille, diamond emblem with too much silver paint, and rear seat immediately under the rear window. Two keyholes. Beaulieu (whatever did Simca say?) without motor. Colors: light green, light gray, light gray and dark blue.

92 GRANDE ÉCHELLE (Voiture de pompiers) 1950

Ladder truck with separate cab, rear body (including hose reel), mount and ladder castings, in the style of the #68 Simplex. The cab can be a conventional (Perche without, Maine with motor) or cabover style (Lyon without, Paris with motor). Color: red.

93 CAMION-BENNE BASCULANTE 1952

Dump truck with cab, rear body and tipper castings. Poitou without, Nivernais with motor. Colors: red, yellow, light green, dark green, yellow and red, light and dark green, light and dark gray.

94 REMORQUE SURBAISSÉE AVEC RAMPE 1952

Flat six-wheel trailer with rear ramp, made to carry a car. America without motor. Colors: red, maroon, yellow-orange, yellow, light green, dark green.

95 VOITURE DE SECOURS 1952

Fire truck with long cab, carrying ten seated firemen on longitudinal benches in its open rear. Lille without, Strasbourg with motor. Color red with yellow benches.

96 VOITURE GRANDE ÉCHELLE COULISSANTE 1952

Fire truck with extending ladder and same cab as #95. Mine may be a blend of the two, as it also has seated firemen. Bordeaux without, Marseille with motor. Color red.

97 VOITURE DE DÉPANNAGE 1952

Wrecker with same cab and rear body as #96, plus a swiveling crane mounted in the back. Nantes without, Limoges with motor. Also exists with seated firemen. Color: red +/- yellow benches.

98 LAMBRETTA SCOOTER 1952

Motor scooter with cast main body, front, and steering front wheel-handlebars parts, dual rear wheels, plastic windshield and driver. The body is more angular than #101. Type A without, B with motor. Colors: metallic light green, metallic green, light gray.

99 LAMBRETTA TRIPORTEUR 1952

Main body of #98 bolted to two-wheel stakeside front body. Type A without, B with motor. Colors: dream, metallic light green, light gray.

100 LAMBRETTA GLACIER 1952

Model #99 with a front superstructure representing a mobile ice cream vendor's stand. Color: cream.

101 TERROT SCOOTER 1952

Motor scooter with front and rear body plus steering front wheel, single rear wheel, plastic windshield and driver; styling is much more rounded than the Lambretta. No motor. Colors: light green, tan, gray, gray-blue, black.

102 CHARIOT ÉLÉVATEUR AVEC GRUE 1952

"Elevator with crane", or the mechanical horse from #104 with a fork-lift equipped with a crane boom. Colors: red, orange, light green, dark green.

103 CHARIOT ÉLÉVATEUR AVEC FOURCHE 1952

This elevator has a normal forklift. Same colors as #102.

104 CHARIOT DE GARE & REMORQUE POUR VÉLOS 1952
"Railway mechanical horse with cycle trailer." The horse is just high enough to hold a motor; the trailer has a tall cycle rack. Colors: red, orange, light green, olive green, light blue horse; silver, red, orange, yellow, light green, light blue.

105 CHARIOT ÉLÉVATEUR AVEC PLATEAU 1952
Back to elevators, this one with a platform instead of a fork. Same colors as #102.

106 CHARIOT ÉLÉVATEUR AVEC BENNE BASCULANTE 1952
This one has a shovel. Same colors as #102.

107 CHARIOT DE GARE AVEC REMORQUES PLATEAU 1952
Another mechanical horse, this one with two flat trailers. Same as #104 minus cycle rack; same colors.

113 FORD COMETE ca. 1952-53
Two-door sedan with single body casting including grille that I once mistook for a Mercedes, three-section rear window, and keyhole on each side. Alençon without motor. Colors: maroon, light blue, metallic blue, light tan and dark blue, light gray and dark green, gray and dark blue, gray and black.

114 PACKARD CABRIOLET ca. 1952-53
Two-door, four-seat convertible with distinctive grille, windshield frame including vent windows, hood ornament and three blobs on each rear fender. Mine has no keyhole. Royan without motor. Colors: white, cream, red, metallic dark red, red-orange, light green, light blue.

115 STUDEBAKER COMMANDER 1953
A nice model of a classic Studebaker with good detail and removable plastic hardtop-windows, just one seat bench inside.
Dinard without motor. Body colors: cream, red, red-orange, yellow, light green, green, light blue, metallic light blue, metallic blue, light tan, black.

116 FORD VEDETTE 1954
A blunt-nosed four-door sedan with its roof painted dark gray and two keyholes. Valence without motor. Colors: cream, red, metallic maroon, dark green, light blue, dark blue, light gray, gray, black, light and dark blue, light gray and blue-gray, gray and dark green.

The following models have two-piece chassis, held on by three bolts, inside the body casting, and including suspension. When they have a motor, it is mounted flat under the hood and wound from below:

191 MERCURY CABRIOLET or HARDTOP 124 mm 1956
Accurate model with body casting, plastic windshield, steering wheel, interior and removable hardtop-windows; casting includes folded top uncovered when hardtop is removed. Colors: red, dark red, maroon, green, turquoise, light blue, blue, metallic blue. Hardtop is light tan, light gray or black.

192 SIMCA VERSAILLES 112 mm 1956
Another convertible, though a less realistic one, with cast body, plastic interior, steering wheel and removable hardtop-windows. Casting does not include folded top, and there is no windshield other than that included in the hardtop. Colors: cream, pink, yellow, light green, green, metallic gray-green, dark blue, black.

193 PEUGEOT 403 112 mm 1956
Four-door sedan with body casting including open roof hatch, plastic interior and steering wheel. Colors: light blue, metallic blue, dark blue, blue-gray, light gray, dark gray, black.

194 MERCEDES-BENZ ? mm 1956
A sedan, quite realistic, with silver grille, bumpers and window frames. Colors: cream, metallic blue, dark blue, light gray, metallic gray.

195 FORD THUNDERBIRD 110 mm 1956
Open roadster with cast body, plastic interior, steering wheel, and removable windshield. A nice model that deserved to be resurrected in the 100 series as #113b. Colors: white, salmon, rose, dark red, orange-red, yellow, light green, turquoise, light blue, light tan, light gray, black. Hardtop is white or black.

196 FORD FAIRLANE BREAK 121 mm 1956
Wagon with cast body, front seat and opening tailgate, plastic steering wheel and removable roof-windows. Removing it turns the car into the #197 pickup truck body. Also exists in white with "Ambulance" decals and plastic patient on stretcher. Colors: white, cream, red, metallic dark red, light green, olive green, metallic olive green, light tan.

197 PICK UP 1956
The Ford Fairlane casting with a removable plastic closed cab. This also exists with a crane like that of #97 mounted at the rear. Same colors as #196.

198 PICK UP BACHÉ 1956
The Ford Fairlane pickup truck with a white plastic rear cover. Same colors as #196.

199 CITROEN DS 19 1956
A very recognizable car with white or black plastic top. Colors: cream, red-orange, light yellow, yellow, light green, green.

207 SIMCA BEAULIEU 1958
A sedan, usually with rear fins in a second color, plus a white plastic top.
Colors: metallic gray green, dark green and metallic gray, dark tan and metallic gray.

208 NOTIN CARAVAN 113 mm 1958
House trailer with cast chassis and axle retainer (as found on many other models), plastic body, opening rear door, and windows with curtains painted on. The body tends to warp. Any car with a tow hook on its rear chassis bolt can pull it. Colors: cream, light green, dark green, light blue, blue.

210 ROLLS-ROYCE SILVER CLOUD 121 mm 1958
Another nice model that reappeared in the 100 series as #115, with cast body and plastic interior. Colors: metallic gray-green, metallic light and dark gray, metallic gray and dark tan.

Numbers unknown:

#? PICKUP TRUCK
The conventional cab sometimes seen on the #92 fire truck and a rear body casting; together they make a very pleasant model that may have been made up from a set but not issued individually.

#? PICKUP TRUCK AND ANIMAL TRAILER 230 MM
A more modern design with cast cab, rear body, opening tailgate and detailed black chassis, plastic windows, interior and steering wheel, pulling a trailer with cast body (= rear body of truck) and chassis-towbar, with plastic stakes, opening tailgate and sheep. The trailer was later used as part of model #612.

The parts in the many Solido sets of this era can be combined to make up various models never offered individually. One example, an impressive heavy truck, will be illustrated.

BABY SERIES

The Baby Series was introduced in or before 1937 and consists of models with single body castings, seats, clockwork motors with keyholes on the left side, metal hubs and rubber tires, except just before and after the war, when one-piece metal wheels were used. Two bolts hold the body to the chassis, which is 90 mm long. Models 117 to 131 were issued before the war; at least some of them were reissued afterward, along with new models, and the series stayed in production until the mid-fifties. No definite information seems to be available, though, as to years of issue.

117 CAMION-CITERNE 1936
Tank truck, looking somewhat like the Junior, with a three-section tank. Colors: chrome, red, dark green, camouflage.

118 PICK-UP 1936
Again there is a certain resemblance to the bigger versions, with an open two-seat cab and rear body. Colors: chrome, red, yellow, light green, green, light blue.

119 COUPÉ SPORT 1936
A sports two-seater with a tailfin of sorts. Colors: chrome, red, light green, light blue, silver blue.

120 BARQUETTE 1936
An open sports-racing car, somewhat similar to #119 and #121 in styling, also with a tailfin. Colors: chrome, red, yellow, light olive green, light blue.

121 CABRIOLET 1936
A four-seat convertible with a folded top at the rear. Colors: chrome, metallic red, maroon, light olive green, dark blue.

122 BERLINE 1936
A four-door sedan this time, but again with similar styling including a grille of vertical lines, rounded at the top and pointed at the bottom. The first type has a much more squarish upper body than the second, and a ridge down the center of the trunk lid that might be called a tailfin. Colors: chrome, red, metallic red, maroon, dark blue.

123 AUTOCAR À CALANDRE VERTICALE 1937

This is a singledeck bus with a wide-V grille of vertical lines and a raised roof section. Colors: chrome, red, yellow, light green. blue.

123 AUTOCAR À CALANDRE HORIZONTALE 1937

The bodywork of this bus is much like the other, but its rounded nose has a grille of horizontal lines. Also listed as #123-A, #123-B, or #124. Colors: chrome, maroon, metallic dark red, light green, light blue, dark blue, tan, camouflage.

124 LIMOUSINE 1937

A fastback four-door sedan with a shield grille of vertical lines and a keyhole in the left rear door. Colors: chrome, red, light green, dark blue, black.

125 LIMOUSINE 1939

A similar fastback four-door design, this time with horizontal grille lines. Colors: chrome, red, yellow, light green, light blue, blue, metallic blue, gray.

126 LIMOUSINE 1950

A sleek fastback "inspired by the Ford Vedette", with a low, wide horizontal grille. Granville without, Deauville with motor. Colors: white, dark red, olive green, light blue, light gray, light metallic gray.

127 FAUX CABRIOLET (Coupé Sport) 1937

The shield radiator with vertical lines returns on this two-door "false convertible" design with rear trunk. Colors: chrome, dark red, light blue, blue.

128 FAUX CABRIOLET (Coupé Sport) 1939

The general design of this second "false convertible" is much like the previous one, but it has a roundish grille with horizontal lines. Colors: chrome, dark red, dark green, light blue, metallic blue.

129 COUPÉ 2 PORTES (Coupé Sport)

This two-door coupé has similar styling to the #126 sedan, and was also inspired by the Ford Vedette. Megève without, Chamonix with motor. Colors: red, dark red, light green, light blue, black.

130 SEMI-REMORQUE BENNE (Tracteur et remorques) 1937

The first title is misleading, as it means "tipping semi-trailer" but the model is actually a truck tractor (prime mover) and a four-wheel open trailer. Tractor colors: chrome, red, orange, army green, blue-violet, black, green camouflage, tan camouflage. Trailer colors: red, yellow, green, blue, blue-violet, tan camouflage. Also available: a two-wheel ammunition trailer in army green, black or camouflage, and a two-wheel camping trailer in yellow or light green.

131 VOITURE AÉRODYNAMIQUE 1937

A long skinny sports-racing coupe with a long pointed tail and flat two-dimensional fenders, plus a grille of vertical lines. Colors: chrome, cream, red, yellow, light green, light metallic green, light blue.

132 COUPÉ SPORT 1950

Another sports-racing coupe, but a much more modern design, with its roof tapering to a point and rounded full-length fenders. Montlhéry without, Le Mans with fenders. Colors: dark red, yellow, light green, light olive green, light blue.

133 PEUGEOT CABRIOLET, FERMÉ 1954

This two-door convertible's top is up and usually painted a second color. The car has a big grille of horizontal lines. Chambery. Colors: light gray, light and dark green, medium and dark green, dark green and light gray, light and dark blue, light and dark gray, light gray and dark blue, light gray and black.

134 AUTOCAR 1950

A squarish, modern singledeck bus with slab front and six windows on each side. Nice without, Monte Carlo with motor. Colors: red, dark red, yellow, light green, olive green, light blue, black.

135-A BERLINE (Studebaker 1947) 1951

A sleek four-door sedan with a grille like an upside-down T. Cabourg without, Biarritz with motor. Colors: white, cream, dark red, maroon, light blue, dark blue, gray.

135-B PEUGEOT 203 BERLINE 1952

A pleasant, sleek fastback sedan with a grille of horizontal lines. Toulouse without, Menton with motor. Colors: metallic red, metallic salmon, light green, metallic blue, silver blue, dark blue, blue-gray, light gray, black.

136 SIMCA ARONDE 1952

A recognizable down-to-earth four-door sedan with simple grille design (no bars except those that outline it); it often has a second-color roof. Annecy without, Bayonne with motor. Colors: red, yellow, turquoise, light blue, metallic light blue, light gray, dark gray, cream and dark green, creaqm and dark blue, light and dark green, light green and dark blue, light and medium blue, light and dark blue, tan and dark blue.

137 NASH LIMOUSINE 1952

A very recognizable four-door Nash, much like the Junior model, with oval grille of vertical lines. Toulon with, Ajaccio without motor. Colors: white, cream, red, maroon, metallic blue, dark blue, light gray, gray.

138 CAMION-CITERNE 1951

A typical cabover design of the fifties, with vertical-line grille and five-section tank. Rouen without, Le Havre with motor. Colors: red, yellow, light green, dark green, blue.

139 FORD CONSUL or ZEPHYR 1954

Another four-door sedan, with distinctive grille and big windows. Avignon. Colors: maroon, light green, dark green, light blue, dark blue, gray, black.

140 KAISER HENRY J 1954

A nice little car, recognizable to those of us who remember the real thing. The wheels and tires of mine are obviously not original. Grenoble. Colors: white, cream, red, metallic red, maroon, light green, light blue-green, sea blue, gray.

141 FOURGON TOLÉ 1951

That means a truck with sheet metal body, and it's either a light van or a van-type ambulance, the latter with red or blue crosses. Neuilly without, Ivry with motor. Colors: white, cream, red, light green.

MOSQUITO SERIES

Here we find the smallest of the old Solido models, each composed of a body casting attached to a common 60 mm chassis, which more or less fits inside it, by only one bolt--not a very practical arrangement even if the bolt is tight, which it usually isn't. The models all have one-piece cast metal wheels, but have no seats or motors. They were all issued in 1952 and stayed in production until 1957.

151 CITROEN 11 CV 66 mm 1952
Recognizable to be sure, but something about it just doesn't look right. The later version with the trunk instead of the spare wheel cover. Colors: red, light green, metallic blue, dark blue, light gray, black.

152 MASERATI SPORT 67 mm 1952
A 2CLT or 4CLT sports coupe that exists in two forms, the first with a very simple grille and side window posts, the second with a more realistic two-section grille of horizontal bars and no window posts. Colors: white, red, metallic red, dark red, yellow, light blue, gray-blue for the first type, white, cream, dark red, light green, blue, gray for the second.

153 RENAULT FRÉGATE 69 mm 1952
A pleasant model of a four-door sedan with characteristic wide grille of horizontal bars. Colors: maroon, light green, olive green, light blue, dark blue, light gray, black.

154 HOTCHKISS ANJOU 69 mm 1952
An unusual model, looking rather like a stately British sedan with an almost-Bugatti horseshoe grille. Colors: metallic rose, dark green, dark blue, metallic purple, gray, black.

155 PEUGEOT 203 BERLINE 67 mm 1952
It's not exactly identical to the Baby version, and there are two types; the first has a cast-in roof hatch and a grille of two horizontal bars; the second has a five-bar grille and apparently lacks a roof hatch. Colors: maroon, metallic maroon, dark green, dark blue, tan, gray-blue, dark gray, black for the first type, white, red, yellow, dark green, dark blue, blue-gray, light gray, dark gray, black for the second.

156 RENAULT PICK UP 67 mm 1952

Some folks say it's a Simca, but Solido says it's a Renault; a very ordinary little pick-up truck. Colors: red, yellow, olive green, dark green, blue-green, light blue, blue-gray, tan, gray, dark gray, black.

157 FORD VEDETTE COACH 68 mm 1952

A typical snub-nosed Vedette with its usual styling and a thin wide grille that gives it a toothy look. Colors: white, red, maroon, light green, light blue-green, tan, light gray.

158 CAMION-CITERNE ? mm 1952

A tank truck rather like the Baby model, with typical cabover styling. Colors: red, dark red, yellow, light green, green. Also red Esso and Mobilgas versions.

159 SIMCA 8 SPORT 70 mm 1952

There are two versions, one with no windshield and a grille of four plain horizontal bars, the other with a windshield frame and a more distinctive grille. Colors: light green, olive green, turquoise, tan, brown for the first type, light green, sky blue, light gray, gray.

160 TATRAPLAN 68 mm 1952

This is a fully streamlined single-seater with a high tailfin-headrest just behind the cockpit, a grille of vertical lines and two rows of louvers on the hood. Colors: red, yellow, light green, green, light blue, sky blue, metallic gray, brown.

161 FORD COMETE 69 mm 1952

A two-door car with the same grille and body style as the #113 Junior model, but without its three-section rear window. Colors: white, olive green, dark green, blue, dark blue, light gray, black.

162 AUTOCAR ? mm 1952

A Latil singledeck bus rather like the #134 Baby model, with six windows on each side. Colors: red, yellow, light green, turquoise, light blue, blue.

Here the early Solido models come to an end. Most of them were out of production by the time the new 100 series was introduced in 1957, but a few were still produced for a year or two afterward. They were simple models, almost primitive by fifties standards, and it was time for them to be replaced by models more in keeping with the trends of the times. The next section of the book will take up the story from the introduction of the new series in 1957.

THE 100 SERIES

They were the right models at the right time. In 1957 the market in France and other European countries was dominated by 1/43 scale diecast models of very good quality, not only in terms of detail and workmanship but also of special features such as clear plastic windows, driver figures, realistic interior details and spring suspension. To gain a significant portion of the market, a new line of products had to be as good as possible in all of these areas and at the same time offer models of interesting vehicles that would attract children and their parents by offering plenty of eye appeal and play value. This is exactly what Solido did.

A few of the last Junior models being made at the same time showed considerable progress in terms of quality and indicated that the potential for producing an outstanding line of products was there, at least in terms of body castings. What remained was to get away from the old type of chassis construction that was based on having to install a windup motor. Very few diecast miniatures of that era had clockwork motors; it seemed to be generally understood that clockwork toys were essentially a different area, involving basically larger toys made to be played with in different ways from 1/43 scale and smaller diecast miniatures that were intended, as the scales involved indicate, for railroad platform use and related uses--setting up toy towns with traffic, running or racing these free-rolling toy vehicles, and the like.

Thus the 100 series came into the world with chassis that harmonized well with their bodies, the two components providing a secure, inconspicuous home for running gear, suspension and interior components. (These and other components, by the way, are subjected to a much greater analysis in Azema's books than is possible or even practical here.) And later it occurred to Solido to put catalog numbers, and even dates of issue, on the chassis along with names, scales and such, which made life simpler for the collector and chronicler.

And what a success they were! To those of us who spent our younger days collecting such brands as Dinky and Corgi, Tekno and Mercury, Solido models were a little bit better--and in some cases more than a little bit. As I look at these models today, in a day when all sorts of special features have become commonplace, I still regard the early 100 series Solido models as something very special. The styles of the vehicles doubtless have something to do with this feeling--let's face it, present-day touring, sporting and commercial vehicles have become so utilitarian that they have lost a good deal of personality--and perhaps our nostalgic feelings for the models we collected when we were young play a role too, but there is just something about these models that makes them look especially good to us, now as then. So let's stop raving and start listing them.

100   JAGUAR D "LE MANS"   92 mm   1/43   1957-1971

Sports-racing car with cast body, gray chassis, plastic windshield, steering wheel and driver, turned hubs, tires, suspension, big or small racing number, sometimes other decals, silver lights.

1. Medium green body, +/- tricolor rondels.
2. Dark green body.
3. Dark red body, +/- Stars & Stripes decal.
4. Orange-red or coral red body.

101   PORSCHE SPYDER   86 mm   1/43   1957-1965

Sports-racing car with cast body, gray chassis, plastic windshield, steering wheel and driver, turned hubs, tires, suspension, big or small racing number decal, green seat cover.

1. Silver body.

102   MASERATI 250 F1   98 mm   1/43   12/1957-1965

Racing car with cast body, gray chassis, unpainted partial exhaust pipes, plastic windshield, steering wheel and driver, turned hubs, tires, suspension, big or small racing number decal, silver grille and cast-in partial exhaust pipes.

1. Dark red body.
2. Brick red body.
3. Light red or orange-red body.

103   FERRARI TYPE 500 TRC   95 mm   1/43   3/1958-1965

Sports-racing car with cast body, gray chassis, plastic windshield, interior details and driver, turned hubs, tires, suspension, big or small racing number decal, silver grille and headlights.

1. Bright red body, large or small number.
2. Dark red body, small number.

104   VANWALL FORMULA 1   101 mm   1/43   5/1958-1968

Racing car with cast body, gray chassis, plastic windshield, steering wheel and driver, turned hubs, tires, suspension, white "Vanwall" and large or small racing number decals, silver cast-in exhaust pipes.

1. Light green body.
2. Dark green body.

105 MERCEDES-BENZ 190SL CABRIOLET   98 mm   1/43   6/1958-1966

Convertible with cast body, gray or black chassis, plastic windshield, interior details and male driver, turned hubs, tires, suspension, black folded top, other painted parts depending on body color.

1. Silver body.
2. Red body.
3. Metallic red body.
4. Metallic gray-green body.
5. Metallic dark blue body.

106 ALFA ROMEO GIULIETTA SPIDER   91 mm   1/43   7/1958-1968

Sports roadster with cast body, gray or black chassis, plastic windshield, interior details and female driver, turned hubs, tires, suspension, silver headlights, grille and bumpers.

1. Gold body.
2. White body.
3. Cream body.
4. Rose body.
5. Red body.
6. Dark red body.
7. Turquoise body.

8. Light blue body.
9. Dark blue body.

107 ASTON MARTIN 3L DBR1 92 mm 1/43 1959-1968

Sports-racing car with cast body, gray or black chassis, plastic windshield, interior details and driver, turned hubs, tires, suspension, big or small racing number decal, painted grille, silver headlights.
   1. Red body, ? grille.
   2. Yellow body, silver grille.
   3. Medium green body, light green grille.
   4. Dark green body, light green grille.

108 PEUGEOT 403 CABRIOLET 103 mm 1/43 1959-1963

Convertible with cast body, gray chassis, plastic windshield, interior details and driver, turned hubs, tires, suspension, black folded top, silver headlights, grille and bumpers.
   1. White body.
   2. Light yellow body.
   3. Reddish tan body.
   4. Olive green body.
   5. Light blue body.
   6. Medium blue body.
   7. Light gray body.

109 RENAULT FLORIDE CABRIOLET 96 mm 1/43 1959-1965

Convertible with cast body, gray chassis, plastic windshield, interior details and female driver, turned hubs, tires, suspension, black folded top, silver headlights and bumpers.
   1. Gold body.
   2. White body.
   3. Rose body.
   4. Red body.

5. Brick red body.
6. Light green body.
7. Light blue body.
8. Dark blue body.
9. Gray body.

110 SIMCA OCEANE CABRIOLET 96 mm 1/43 10/1959-1963

Convertible with cast body, gray chassis, plastic windshield, interior details and female driver, turned hubs, tires, suspension, black folded top, silver headlights, grille and bumpers.
   1. Rose body.
   2. Light red or orange-red body.
   3. Light green body.
   4. Dark green body.
   5. Greenish blue body.
   6. Blue body.
   7. Dark blue body.

111 ASTON MARTIN DB4 105 mm 1/43 10.1960-1965

Sports coupe with cast body, gray chassis, plastic windows and interior, turned hubs, tires, suspension, racing number and stripe decals (usually), silver headlights, grille and bumpers.
   1. Cream body.
   2. Red body.
   3. Orange body.
   4. Light yellow body.
   5. Light green body.
   6. Dark green body.
   7. Light metallic gray body, no decals.
   8. Dark metallic gray body.
   9. Dark metallic gray body, no decals.

112 DB-PANHARD LE MANS 84 mm 1/43 11/1960-1971

Sports-racing car with cast body, gray chassis, plastic windshield, dash and driver, turned hubs, tires, suspension, racing stripe and number decals, silver headlights, red taillights and seats.
   1. Blue body.

113 FIAT-ABARTH 99 mm 1/43 1961-1970

Speed record car with cast body, opening hatch, gray chassis, plastic windshield and steering wheel, turned hubs, tires, suspension, "Fiat", "Abarth" and emblem decals, silver lights if body color permits.
   1. Silver body.
   2. Red body.
   3. Dark red body.
   4. Orange or orange-red body.
   5. Metallic blue body.

113-B FORD THUNDERBIRD 109 MM 1/43 1959-1963

Roadster with cast body (= Junior #195), gray chassis, plastic windshield, interior details and male driver, turned hubs, tires, suspension, silver headlights, grille and bumpers.
   1. White body.
   2. Rose body.
   3. Red body.
   4. Orange-red body.
   5. Light green body.
   6. Turquoise body.
   7. Light blue body.
   8. Metallic dark blue body.

114 CITROEN AMI-6 87 mm 1/43 10/1961-1965

Sedan with cast body, gray chassis, plastic windows and interior details, turned hubs, tires, suspension, wire front bumper, silver rear bumper, red taillights, sometimes white roof.
   1. White body.
   2. Ivory body.
   3. Light green body, white roof.
   4. Green body.
   5. Sea blue body.
   6. Sky blue body, white roof.
   7. Blue-gray body.
   8. Light gray body.

115 ROLLS-ROYCE SILVER CLOUD 121 mm 1/43 1960-1967

Sedan with cast body (= Junior #210), gray or black chassis, plastic windows and interior details, turned hubs, tires, suspension, silver grille and bumpers, gold headlights, red taillights. Upper body color includes roof, hood and trunk lid.
   1. Gold upper, metallic red-brown lower body.
   2. Silver upper, metallic maroon lower body.
   3. Silver upper, dark olive green lower body.
   4. Silver upper, metallic gray lower body.
   5. Metallic gray-green entire body.

116 COOPER 1500 84 mm 1/43 10/1960-1968

Racing car with cast body, gray or black chassis, unpainted exhaust pipes, plastic windshield, steering wheel and driver, turned hubs, tires, suspension, racing number and sometimes stripe decals, painted nose if no stripe decal.
   1. Silver body, white nose.
   2. Light yellow body, silver nose.
   3. Yellow body, red nose.
   4. Pale green body, stripe decals.
   5. Green body, yellow nose.
   6. Dark green body, yellow nose.

7. Blue body, stripe decals.
8. Metallic blue-gray body, yellow nose or stripe decals.
9. Black body, white nose or stripe decals.

117 PORSCHE FORMULA II 81 mm 1/43 3/1961-1967

Racing car with cast body, gray or silver chassis, plastic windshield, steering wheel and driver, turned hubs, tires, suspension, racing number decal.
1. Silver body.
2. Yellow body.
3. Light green body.
4. Black body.

118 LOTUS FORMULA I 77 mm 1/43 1961-1968

Racing car with cast body, gray or black chassis, plastic windshield, steering wheel and driver, turned hubs, tires, suspension, racing number and sometimes stripe decals.
1. Light green body.
2. Dark green body.
3. Dark green body, white stripes.

119 CHAUSSON BUS 142 mm 1/43 1961-1967

Singledeck bus with cast body, gray or unpainted base (bolted or riveted on), later roof rack, +/- plastic windows, turned hubs, tires, +/- suspension, +/- wire antenna with plastic or metal top, +/- silver trim, +/- decals. See Azema for more data.

A. City bus, no roof rack or antenna, cream or light gray trim, silver bottom panels:
1. Dark red body.
2. Wine red body.
3. Yellow body.

4. Bright green body.
5. Metallic green body.
6. Almond green body.
7. Metallic blue body.
8. Purplish blue body.

B. Interurban bus with roof rack, silver bottom panels:
1. Light blue body, Air France and Orly decals.
2. Light blue body, RTF and Télévision Française Car de Reportage decals, +/- antenna.
3. Light blue body, no decals, +/- antenna.
4. Dark blue body, Air France and Orly decals.
5. Dark blue body, Bordeaux and Citram decals.

120 CHAUSSON TROLLEYBUS 142 mm 1/43 1961-1967

Singledeck bus with cast body (= 119), gray chassis, cast or plastic roof cap, cast or plastic trolley poles, plastic windows, turned hubs, tires, +/- suspension, silver bottom panels.
1. Dark red body.
2. Wine red body.
3. Bottle green body.
4. Dark green body.

121 LANCIA FLAMINIA COUPE 108 mm 1/43 6/1961-1967

Coupe with cast body, opening doors, gray or black chassis, plastic windows and interior, turned hubs, tires, suspension, silver headlights, grille and bumpers, red or orange taillights.
1. Cream body.
2. Red body.
3. Metallic red body.
4. Metallic light red body.
5. Metallic wine red body.
6. Pale green body.

7. Metallic gray-green body.
8. Light gray body.
9. Metallic gray body.
10. Metallic dark gray body.
11. Metallic brownish body.

122   FERRARI FORMULA I   98 mm   1/43   2/1962-1968

Racing car with cast body, gray chassis, motor, plastic windshield, steering wheel and driver, metal roll bar, turned hubs, tires, suspension, number and sometimes stripe decals.
    1. Dark red body.
    2. Dark red body, gold stripes.
    3. Dark red body, white stripes.

123   FERRARI 250 GT 2+2   105 mm   1/43   1962-1968

Coupe with cast body, opening doors, gray or silver chassis, plastic windows and interior, spoked hubs, tires, suspension; many variations: see Azema.
    1. Gold body.
    2. White body.
    3. Ivory or cream body.
    4. Red body.
    5. Metallic red body.
    6. Dark red body.
    7. Wine red body.
    8. Metallic wine red body.
    9. Metallic light green body.
    10. Metallic green body.
    11. Turquoise body.
    12. Light blue body.
    13. Sky blue body.
    14. Dark blue body.

15. Metallic dark blue body.
16. Metallic blue-gray body.
17. Metallic light gray body.
18. Metallic gray body.
19. Metallic dark gray body.
20. Metallic gold-gray body.
21. Metallic bronze body.

124   ABARTH 1000   80 mm   1/43   9/1962-1969

Sports fastback with cast body, opening doors, bray or black chassis (first to have catalog number and date of issue), plastic windows and interior, turned hubs, tires, suspension, racing number +/- "Fiat Abarth 1000" or emblem and stripe decals.
    1. Silver body.
    2. Red body.
    3. Dark red body.
    4. Metallic dark blue body.
    5. Light gray body.

125   ALFA ROMEO 2600   103 mm   1/43   3/1963-1967

Coupe with cast body, opening doors, gray chassis, plastic windows and interior, turned hubs, tires, suspension, silver headlights, grille and bumpers, red taillights.
    1. White body.
    2. Bluish-white body.
    3. Red body.
    4. Metallic red body.
    5. Dark red body.
    6. Light blue body.
    7. Metallic blue body.
    8. Light gray body.
    9. Metallic gray body.

10. Metallic dark gray body.
11. Brown body.
12. Metallic lilac brown body.
Note: Other colors used in sets.

126 MERCEDES-BENZ 220 SE 111 mm 1/43 3/1963-1968
Two-door sedan with cast body, opening doors, silver chassis, plastic windows and interior, turned hubs, tires, suspension, silver headlights, grille, bumpers and bottom panels, red taillights.
1. Red body.
2. Metallic red body.
3. Metallic wine-red body.
4. Light green body.
5. Metallic dark green body.
6. Metallic blue body.
7. Metallic blue-black body.
8. Light gray body.

127 NSU PRINZ IV 79 mm 1/43 4/1963-1967
Two-door sedan with cast body, opening doors, gray or black chassis, plastic windows and interior, turned hubs, tires, suspension, gold headlights, red taillights, silver bumpers.
1. Silver body.
2. Red body.
3. Dark red body.
4. Metallic reddish tan body.
5. Light green body.
6. Blue body.
7. Light gray body.

128 FORD THUNDERBIRD 121 mm 1/43 5/1963-1967
Coupe woth cast body, opening doors, gray or black chassis, plastic windows and interior, turned hubs, tires, suspension, silver headlights, grille and bumpers, red taillights.
1. Red body.
2. Brick red body.
3. Maroon body.
4. Yellow body.
5. Light green body.
6. Green body.
7. Turquoise body.
8. Light blue body.
9. Metallic light blue body.
10. Metallic dark blue body.
11. Metallic tan body.
12. Light gray body.
13. Metallic gray body.
14. Metallic olive gray body.
15. Metallic gray-brown body.

129 FERRARI 2.5 LITER 94 mm 1/43 2/1964-1969
Sports-racing car with cast body, gray or black chassis, plastic windshield, interior and driver, turned hubs, tires, suspension, silver headlights and grille, #152 and Ferrari emblem decals.
1. Red body.
2. Orange-red body.
3. Dark red body.

130 ASTON MARTIN DB5 VANTAGE 105 mm 1/43 3/1964-1968
Sports coupe with cast body (=#111), opening doors, gray or black chassis, plastic windows and interior, spoked hubs, tires, suspension, silver headlights, grille and bumpers. Note: #111 has no opening doors, no "Vantage" on chassis.
1. Silver body.

2. Red body.
3. Metallic red body.
4. Yellow body.
5. Metallic light green body.
6. Green body.
7. Metallic green body.
8. Turquoise body.
9. Metallic dark blue body.
10. Metallic dark blue body.
11. Metallic tan body.

131 B.R.M. V8 87 mm 1/43 4/1964-1970
Racing car with cast body, gray chassis, stack exhaust pipes, plastic windshield, steering wheel and driver, metal roll bar, turned hubs, tires, suspension, racing number and sometimes stripe decals.
1. Yellow body.
2. Light green body.
3. Dark green body.
4. Blue body.
5. Pale bluish-gray body.
Note: Spoked (as #170) or racing (as #152) wheels are also found on dark green models.

132 MERCEDES SS 1928 111 mm 1/43 1/1964-1980>#4132
Tourer, top up, with cast body, opening doors, black chassis, plastic interior, black smooth or textured top, windshield with silver frame, silver grille, exhaust pipes, lights and bumpers, spoked hubs, treaded tires, side spares, suspension, hood strap.
1. White or ivory body.
2. Red body.
3. Metallic red body.
4. Pale green body.

5. Light blue body.
6. Metallic gray body.

133 FIAT 2300/S GHIA CABRIOLET 105 mm 1/43 6-1964-1969
Open convertible with cast body, opening doors, gray or black chassis, plastic windows and interior, cast hubs, tires, suspension, jewel headlights, silver grille and bumpers.
1. White body.
2. Cream body.
3. Red body.
4. Metallic red body.
5. Dark red body.
6. Metallic wine red body.
7. Light blue body.
8. Metallic blue body.
9. Tan body.
10. Pale gray-green body.
11. Metallic gray body.

134 PORSCHE G.T. LE MANS 93 mm 1/43 5/1964-1971
Sports-racing coupe with cast body, opening doors, matching chassis, plastic windows and interior, turned or cast hubs, tires, suspension, gold or silver headlights, red taillights. Note: Also exists with racing (as #192) hubs.
1. Silver body and chassis.
2. Ivory body and chassis.
3. Metallic light gray body.
4. Pale gray body.

135 LOLA-CLIMAX V8 F.I 89 mm 1/43 1/1965-1966
Racing car with cast body, silver or black chassis, plastic windshield, steering wheel, driver, silver air scoop, exhaust pipes and suspension,

turned or cast hubs, tires, suspension, racing number and sometimes nose trim decals.

1. Chromed body.
2. Aqua body.
3. Green body.
4. Dark green body.
5. Metallic light blue body.
6. Metallic dark blue body.

136 BUGATTI 41 ROYALE 1930 135 mm 1/43 8/1964-1980>#4136

Town car with cast body, removable hood, opening front doors, matching chassis, unpainted motor and hubs, plastic windshield, windows and interior, silver and black radiator, silver lights, bumper and parts, black tires, suspension.

1. White body.
2. Metallic wine red body.
3. Metallic dark green body.
4. Metallic dark blue body.
5. Metallic dark gray body.
6. Black body.

137 MERCEDES SS 1928 113 mm 1/43 1/1965-1980>#4137

Tourer, top down, with castings and parts as #132 but with folded windshield and top.

1. White body, black chassis.
2. Silver body, black chassis.
3. Metallic red body, black chassis.
4. Light metallic gray body, dark metallic gray chassis.
5. Light metallic gray body, black chassis.
6. Dark metallic gray body, light metallic gray chassis.
7. Golden tan body, black chassis.

138 HARVEY ALMUMINUM SPECIAL 79 mm 1/43 4/1965-1971

Racing car with cast body, matching chassis, plastic windshield, steering wheel and driver, silver air intakes, roll bar and exhaust pipes, racing wheels, tires, suspension, decals including name, trim and #82.

1. Red body.
2. Orange-red body.
3. Dark red body.
4. Dark green body.
5. Sky blue body.

139 MASERATI 3.5 102 mm 1/43 4/1965-1971

Sports coupe with cast body, matching opening hood and doors, black chassis, plastic windows, red or black interior, silver bumpers and engine, jewel lights, spoked hubs, tires, suspension.

1. Red body.
2. Metallic red body.
3. Metallic orange-red body.
4. Maroon body.
5. Metallic gray-green body.
6. Metallic light blue body.
7. Brown body.
8. Metallic silver gray body.

140 PANHARD-LEVASSOR 1925 120 mm 1/43 7/1965-1980>#4140

Landaulet with cast body, matching opening front doors, black chassis, plastic windows, red or tan (or black?) interior, black top, grille and twin spares, gold radiator shell, windshield frame and lights, brown trunk, spoked hubs, tires, suspension.

1. Metallic red body.
2. Metallic creamy red body.
3. Metallic red-orange body.
4. Metallic dark red body.

5. Metallic wine-red body.
6. Metallic green-gold body.
7. Metallic dark green body.
8. Metallic dark blue body.
9. Purple body.
10. Metallic tan body.
11. Metallic grayish body.

141 CITROEN AMI-6 88 mm 1/43 4/1965-1968
Wagon with cast body, matching opening hatch, black chassis, plastic windows, red or tan interior, wire bumper, silver bumpers, grille and headlights, cast hubs, tires, suspension.
Also exists with newer wheels and tires, and with trailer hitch.
1. Cream body.
2. Wine-red body.
3. Green body.
4. Light blue body.
5. Dark blue body.
6. Metallic dark blue body.
7. Dark gray body.

142 ALPINE FORMULA III 84 mm 1/43 7/1965-1969
Racer with cast upper and lower body, plastic windshield, driver, silver roll bar, mirrors, exhaust pipe and suspension, cast racing wheels, tires, suspension, black number on white disc.
1. Blue upper and lower body, number 1 to 10.

143 PANHARD 24 BT 105 mm 1/43 2/1966-1968
Two-door with cast body, matching opening hood, doors and trunk, black chassis, plastic windows and headlights, red, tan or black interior, engine, silver bumpers and hubs, tires, suspension.
1. White body.
2. Silver body.

3. Rose red body.
4. Orange-red body.
5. Maroon body.
6. Dark green body.
7. Light blue body.
8. Purple body.

144 VOISIN CARENE 1934 106 mm 1/43 2/1966-1980>#4144
Vintage car with cast body, chassis, +/- opening doors, drive train, exhaust pipe, plastic windows, red or ? interior, silver lights and radiator shell, black grille, cast or plastic hubs, tires, two rear spares, suspension.
1. Cream body and chassis.
2. Cream body, black chassis.
3. Pale yellow body, black chassis.
4. Light metallic blue body and chassis.
5. Metallic blue body and chassis.
6. Dark blue body and chassis.
7. Metallic dark blue body and chassis.
8. Light gray body and chassis.
9. Black body and chassis.

145 HISPANO-SUIZA H6B 1926 115 mm 1/43 7/1966-1980>#4145
Victoria with cast body, chassis, drive train, exhaust pipe, plastic windshield, brown interior, black top, trunk and grille, silver radiator shell, lights, windshield frame and irons, cast or plastic hubs, tires, twin spares, suspension.
1. Cream body, dark maroon chassis.
2. Wine-red body, vanilla chassis.
3. Wine-red body, black chassis.
4. Light green body, dark brown chassis.
5. Light green body, black chassis.
6. Metallic dark green body, metallic gray chassis.

146 FORD GT 40 LE MANS 94 mm 1/43 5/1966-1970
Sports-racing coupe with cast body, matching opening rearblack chassis, engine, plastic windows, red interior, blue hood and stripe decals, number 1, 2 or 12, spoked hubs, tires, suspension.
1. White body.
2. White body, #5272 cast into rear.

147 FORD MUSTANG 106 mm 1/43 5/1966-1970
Sports coupe with cast body, matching opening doors, black chassis, plastic windows, red, white or black interior, jewel lights, silver grille, bumpers and hubs, tires, suspension. Early issues had plastic door liners, later issues did not.
1. Ivory body.
2. Bright red body.
3. Orange-red body.
4. Light blue body.
5. Metallic gray body.
6. Wine-red and black body.

147b FORD MUSTANG RALLY 106 mm 1/43 7/1966-1972
Sports coupe with castings and parts as #147 plus black plastic lights, rally decals.
1. White body, Monte Carlo and #180 decals.
2. Yellow body, Monte Carlo and #180 decals.
3. Metallic silver gray body, Monte Carlo and #180 decals.
4. White body, Spa-Sofia-Liege and #9 decals.
5. Yellow body, Spa-Sofia-Liege and #9 decals.
6. Orange body, Spa-Sofia-Liege and #9 decals.

148 ALFA ROMEO GIULIA TZ 90 mm 1/43 6/1966-1972
Sports coupe with cast body, matching opening doors, black chassis, plastic windows and headlights, black interior, red-white stripe and black number

decals (2 types), cast hubs, tires, suspension.
1. Red body, Criterium Cevennes and #82 decals.
2. Orange-red body, decals as above.
3. Dark red body, decals as above.
4. Metallic silver gray body, decals as above.
5. Orange-red body, #2 and hood stripe decals only.
6. Metallic silver gray body, decals as on type 5.

149 RENAULT 40 CV 1926 1/1967-1980>#4149
Landaulet with cast body, matching opening hood, black chassis, drive train, plastic windows, gray seats, black rear body and trunk, silver engine, gold windshield frame, lights and spoked hubs, tires, suspension.
1. Wine-red body.
2. Metallic wine-red body.
3. Maroon body.
4. Dark green body.
5. Dark blue body.
6. Light gray body.

150 OLDSMOBILE TORONADO 123 mm 1/43 3/1967-1973
Two-door hardtop with cast body, matching opening doors, black chassis with battery hatch, plastic windows, white or black interior, silver bumpers and hubs, tires, suspension.
1. Metallic gold body.
2. Metallic green-gold body.
3. Yellow body.
4. Metallic orange body.
5. Wine-red body.
6. Metallic wine-red body.

151 PORSCHE CARRERA 6 94 mm 1/43 5/1967-1972
Sports-racing coupe with cast body and chassis, plastic windows and

headlights, opening gullwing doors, black interior, cast racing hubs, tires, number and trim decals, suspension.

1. Silver body and chassis, #31 or #37 and red front hood decals.
2. Red body and chassis, #18 and white stripe decals.
3. Red body, red or silver chassis?, #37 and front hood decals.
4. Yellow body and chassis, #31 and red front hood decals.
5. Dark yellow body, silver chassis, #37 and red front hood decals.
6. Pale bluish body and chassis, #31 and red front hood decals.
7. Pale bluish body, silver chassis?, #37 and red front hood decals.

152 FERRARI 330 P3 96 mm 1/43 6/1967-1971
Sports-racing coupe with cast body, chassis, matching opening rear hood, plastic windows and headlights, silver or chrome engine and cover, black interior, wire hood supports, cast hubs, tires, spare wheel, number, P or F, and emblem decals.

1. Red body and chassis, #14 or #21 and P decals.
2. Red body and chassis, #14 and F decals.
3. Red-orange body and chassis, #14 and F decals.

153 CHAPPARAL 2D 93 mm 1/43 12/1967-1972
Sports-racing coupe with cast body, matching chassis, air scoop and opening rear hood, unpainted engine, plastic windows and headlights, opening gullwing doors, black interior, spoked hubs (3 types), tires, suspension, name and number 7 decals.

1. White body.
2. Cream body.

154 FIAT 525N 1929 106 mm 7/1967-1980>#4154
Open vintage car with cast body, chassis and matching opening front doors, plastic windows, light gray interior, black folded top and grille, silver radiator shell, lights, bumpers and hubs, tires, twin spares, suspension, three figures, including Pope.

1. Cream body and chassis.
2. Light blue body and chassis.
3. Blue body, black chassis.
4. Metallic blue body, black chassis, no figures.
5. Dark blue body and chassis.
6. Metallic purplish-blue body, black chassis, no figures.
7. Dark gray body and chassis.
8. Black body and chassis.

155 FERRARI SUPERFAST (announced 1967, never made)

156 DUESENBERG J 1931 128 mm 1/43 4/1969-1980>#4156
Limousine with cast body, matching opening front doors, black chassis, plastic windows, black top, gray interior, silver bumpers, grille and lights, spoked hubs, whitewall (or chromewall) tires, twin side spares, suspension.

1. Red body.
2. Orange-red body.
3. Orange body.
4. Light blue body.
5. Mauve body.
6. Tan body.
7. Light gray body.

157 BMW 2000 CS 103 mm 1/43 2/1968-1971
Two-door hardtop with cast body, matching opening hood, doors and trunk, and black chassis, plastic windows and lights, black, red or purple? interior, silver engine, grille and bumpers, chromed or plain hubs, tires, suspension.

1. Silver body.
2. White body.
3. Cream body.
4. Turquoise body.

157b BMW 2000 CS RALLY 103 mm 1/43 10/1968-1974
Castings and parts as #157 with different hood color, plain or chromed lights, tinted windows, #71 and rally decals.
1. Silver body, black hood, Monte Carlo Rally decal.
2. White body, black hood, decals as above.
3. Red body, black hood, decals as above.
4. Orange body, white hood, decals as above.
5. Yellow body, white hood, decals as above.
6. Orange body, white hood, no hood decal, BMW emblem on trunk, Shell and Wynn's decals on sides.

158 ALFA ROMEO AND HOUSE TRAILER ___ mm 1/43 11/1967-1970
#106 Alfa Romeo Giulietta Spider with tow hook, towing cream Junior #208 Nottin house trailer with light blue chassis.
1. White car body.
2. Red car body.
3. Orange car body.

159 CITROEN AMI-6 AND BOAT TRAILER __ mm 1/43 11/1967-1970
#141 Citroen Ami-6 wagon with tow hook, towing light blue trailer with red and white boat (red hull/white deck or white hull/red deck).
1. Wine-red car body.
2. Light blue car body.

160 FORD THUNDERBIRD AND CARAVAN (announced 1967, not issued)

161 LAMBORGHINI MIURA 99 mm 1/43 4/1968-1974
Sports coupe with cast body, matching opening front and rear hoods and doors, chassis, unpainted motor, plastic windows and lights, black or white interior, black louvers, grilles and bumper, plain or silver hubs (several types), tires, suspension.
1. Gold body, silver chassis.
2. Silver body and chassis.
3. Silver body, red chassis, black interior.
4. Silver body, orange chassis.
5. Red body, silver chassis.
6. Orange-red body, silver chassis.
7. Orange-red body, light gray chassis, black interior.
8. Coral body.
9. Yellow body, silver chassis, black interior.
10. Metallic yellow-green body.
11. Turquoise body, light gray chassis, white interior.

162 and 163: numbers never used.

164 SIMCA 1100 91 mm 1/43 9/1967-1974
Sedan with cast body, matching opening front doors and hood, plastic windows, jewel headlights, red or white interior, black chassis, silver engine, grille, bumpers and hubs, tires, suspension.
1. White body, red interior.
2. Brick red body.
3. Maroon body, white interior.
4. Metallic green body.
5. Metallic silver blue body, white interior.

164b SIMCA 1100 POLICE CAR (never issued)

165 FERRARI 365 GTB4 102 mm 1/43 12/1969-1973
Sports coupe with cast body, matching opening doors and hood, plastic windows and lights, black chassis-interior, black and silver engine, silver bumpers, star hubs, tires, suspension.

1. Red body.
2. Metallic dark red body.
3. Metallic green-gold body.
4. Blue body.
5. Metallic tan body.
6. Metallic gray body.

166   DE TOMASO MANGUSTA   96 mm   1/43   4/1969-1976

Sports coupe with cast body, matching opening doors and rear hoods, plastic windows and lights, black chassis-interior and grille, silver engine, spoked hubs, tires, suspension.
1. Gold body.
2. Silver body.
3. Metallic dark red body.
4. Orange-red body.
5. Metallic orange body.
6. Green-gold body.
7. Red body, black rear hoods.

167   FERRARI FORMULA I   95 mm   1/43   7/1968-1971

Racing car with cast body, plain engine, plastic windshield, yellow seat, light gray dash and roll bar, white or silver exhaust pipes, silver suspension, star hubs, tires, Ferrari emblem and #18 or #14 decals.
1. Bright red body, silver exhausts, #18.
2. Orange-red body, details as above.
3. Bright red body, front fins, white exhausts, #14.

167b FERRARI FORMULA 1 WITH AIRFOIL 95 mm 1/43 7/1969-1972

Casting with front fins, parts as above, black dash and roll bar, white exhausts, plus red plastic airfoil on wire mount.

1. Red body, #26 decal only.
2. Darker red body, no decals.

168   ALPINE-RENAULT 3 LITER   108 mm   1/43   5/1969-1976

Sports-racing coupe with cast body, matching opening rear hood, plastic windows and lights, black chassis-interior, silver engine, numerous decals including #30.
1. Blue body, no roof decal.
2. Light blue body, decals as above.
3. Metallic dark blue body, Renault decal on roof.

169   CHAPPARAL 2F   88 mm   1/43   6/1968-1974

Sports-racing coupe with cast body, black chassis, plastic windows and lights, white or cream airfoil and opening doors (2 types), black interior and grille, #4, 5 or 7 decal, spoked hubs, tires, suspension.
1. White body.
2. Cream body.

170   FORD MARK IV   100 mm   1/43   2/1969-1974

Sports-racing coupe with cast body, matching chassis and opening rear hood, plain engine, plastic windows and lights, black or gray interior, black wiper, spoked hubs, silver mirrors, black or white exhausts, racing number and stripe decals (various types).
1. Gold body.
2. Silver body.
3. White body.
4. Red body.
5. Metallic red body.
6. Metallic orange-gold body.
7. Yellow body.
8. Dark blue body.

171 OPEL GT 94 mm 1/43 2/1969-1977

Sports coupe with cast body, matching opening doors and headlight covers, plastic windows, jewel lights, black chassis-interior, silver bumpers, plain hubs, tires, suspension.
1. Red body.
2. Yellow body.
3. Metallic green-gold body.
4. Metallic green body.
5. Blue body.
6. Metallic blue body.
7. Metallic grayish-tan body.
8. White body, black hood and stripes, Greder Racing decals.

172 ALFA ROMEO CARABO BERTONE 94 mm 1/43 6/1969-1977

Futuristic coupe with cast body, mathcing opening doors, black chassis, tinted plastic windows, black interior and louvers, plain hubs, tires, suspension, orange nose stripe.
1. Red and gray body.
2. Red and black body.
3. Metallic green body.
4. Brownish-green body.

173 MATRA FORMULA I 103 mm 1/43 6/1969-1974

Racing car with cast body, plain engine, plastic windshield, driver, blue airfoil, silver suspension, silver or black exhaust pipes, spoked hubs, tires, decals including #5 or #8.
1. Blue body, #8.
2. Light blue body, #5 or #8.

174 PORSCHE 908 LE MANS 108 mm 1/43 12/1969-1976

Sports-racing coupe with cast body, matching opening doors, plastic windows and lights, white airfoil on red brackets, black chassis-interior, 5-spoke hubs, tires, suspension.
1. White body.

175 LOLA T70 MK 3B 95 mm 1/43 1/1970-1976

Sports-racing coupe with cast body, matching chassis and opening doors, plastic windows and lights, black interior, silver engine, 5-spoke hubs, tires, suspension, #6 and other decals.
1. Red body.
2. Metallic blue body.
3. Metallic greenish-blue body.

176 McLAREN M8B CAN-AM 94 mm 1/43 3/1970-1976

Sports-racing car with cast body, plastic windshield, white airfoil, black chassis-interior, silver engine and mirrors, 4-spoke hubs, tires, decals including #4.
1. Orange body.
2. Yellow body.

177 FERRARI 312P 96 mm 1/43 6/1970-1975

Sports-racing coupe with cast body, matching opening rear hood, plastic windows and lights, black chassis-interior and louvers, white exhaust pipes, star hubs, tires, suspension, decals including #18.
1. Red body.

178 MATRA 650 95 mm 1/43 7/1970-1974

Sports-racing car with cast body, matching opening rear hood, black chassis, plastic windshield and lights, white cockpit cover, black seats, silver engine, roll bar and mirror, 6-spoke hubs, tires, suspension, decals including #10.
1. Blue body.

179 PORSCHE 914/6 90 mm 1/43 7/1970-1975

VW-Porsche coupe with cast body, matching opening trunk and headlight covers, plastic windshield, jewel lights, black top and chassis-interior, silver bumpers, spoked hubs, tires, suspension.

1. Red body.
2. Orange body.
3. Yellow body.

180 MERCEDES-BENZ C-111 101 mm 1/43 11/1970-1974

Sports coupe with cast body, matching or black opening rear hood, black chassis, plastic windows, red or black interior, black grilles, silver engine, star hubs, tires, suspension.

1. Metallic silver body, red interior.
2. Metallic red body, black interior.
3. Orange body, black interior.
4. Orange-brown body, black interior.
5. Metallic silver body, black hood, red interior?
6. Red body, black hood and interior.
7. Metallic red body, black hood and interior.
8. Orange body, black hood and interior.

181 ALPINE-RENAULT 1600 90 mm 1/43 11/1970-1980>#1181

Sports coupe with cast body, matching opening doors, clear or tinted plastic windows and lights, black chassis-interior, silver bumpers, 6-spoke hubs, tires, suspension, various decals, some versions have black plastic lights on front bumper.

1. White body, #24, red trim and Defense Mondiale decals.
2. Blue body, #18 and Monte Carlo 1970 decals, bumper lights.
3. Metallic light blue body, otherwise as type 2.
4. Metallic dark blue body, otherwise as type 2.
5. Blue body, #18 and Monte Carlo 1973 decals, bumper lights.
6. Metallic light blue body, otherwise as type 5.

7. Metallic pale blue body, no bumper lights, otherwise as type 5
8. Blue body, #8 and BSB decals, no bumper lights; special issue.

182 FERRARI 572S 92 mm 1/43 10/1970-1976

Sports-racing coupe with cast body, matching opening doors, silver chassis, plastic windows and lights, black interior, white engine cover, star hubs, tires, suspension, decals including #28.

1. Red body.
2. Yellow body.
3. Metallic gray body.

183 ALFA ROMEO ZAGATO JUNIOR 88 mm 1/43 9/1971-1979

Coupe with cast body, matching opening doors, palstic windows, jewel lights, black chassis-interior and bumpers, plain hubs, tires, suspension, green 4-leaf clover decals.

1. Ivory body.
2. Red body.
3. Metallic red body.
4. Yellow body.
5. Metallic yellow-gold body.
6. Metallic blue body.
7. Red-brown body.

184 CITROEN SM 112 mm 1/43 12/1970-1978

Two-door with cast body, matching opening doors, plastic windows and lights, black chassis, gray or tan interior, silver bumpers and hubs, tires, suspension. Headrests not always color of seats.

1. Red body.
2. Rose-gold body.
3. Metallic salmon body.
4. Light green body.
5. Metallic light blue body.

6. Metallic blue body.
7. Metallic dark blue body.
8. Metallic gray-green body.
9. Metallic gray-brown body.
10. Metallic bronze body.

185 MASERATI INDY 108 mm 1/43 3/1971-1977

Sports coupe with cast body, matching opening doors, plastic windows, black chassis-interior, silver bumpers and hubs, tires, suspension.
1. Yellow gold body.
2. Rose gold body.
3. Metallic salmon body.
4. Metallic wine red body.
5. Metallic orange body.
6. Yellow body.
7. Metallic green body.
8. Metallic turquoise body.
9. Metallic blue body.
10. Metallic blue-gray body.
11. Metallic light gray body.
12. Metallic gray body.
13. Metallic bronze body.

186 PORSCHE 917 98 mm 1/43 4/1971-1976

Sports-racing coupe with cast body, matching opening rear hood, plastic windows and lights, black chassis-interior, silver motor, cream fan, various decals.
1. Orange-red body, #23, Shell and white stripe decals.
2. Bright red body, decals as above.
3. Light blue body, #2 and Gulf decals.

186M PORSCHE 917 MARTINI 98 mm 1/43 11/1971-1980

Same basic model as #186, with #38, Martini and red-blue-purple stripe decals.
1. Silver body.

187 ALFA ROMEO 33/3 102 mm 1/43 5/1971-1975

Sports-racing car with cast body, matching opening rear hood, plastic windshield and lights, red airfoil and mirror, black chassis-interior, silver motor, roll bar and hubs, tires, suspension, #37 and Alfa Romeo decals.
1. Red body.

188 OPEL MANTA 99 mm 1/43 6/1971-1976

Coupe with cast body, matching opening doors, plastic windows and lights, black chassis-interior and grille, silver bumpers and hubs, tires, suspension.
1. Red body, black hood.
2. Orange body, black hood.
3. Metallic olive body.
4. Metallic blue-green body.
5. Metallic gray body.

188R OPEL MANTA RALLY 99 mm 1/43 2/1976-1978

Same basic model as #188, with #43, Esso and other decals.
1. White body.

189 BERTONE BUGGY 82 mm 1/43 7/1971-1976

Beach buggy with cast body, plastic windshield, black chassis-interior, silver roll bar, bumpers and hubs, tires, suspension, black trim decals.
1. Red-orange body.
2. Yellow body.
3. Yellow gold body.
4. Light green body.
5. Metallic green body.

190 FORD CAPRI 2900 97 mm 1/43 1/1972-1975

Coupe with cast body, matching opening doors, tinted plastic windows, black chassis-interior and grille, jewel lights, spoked hubs, tires, suspension, #22, Castrol and other decals.
1. White to cream body, blue hood.
2. Yellow body, black hood.

192 ALPINE-RENAULT A310 96 mm 1/43 2/1972-1980>#1192

Sports coupe with cast body, matching opening doors, plastic windows, opening hatch, clear or tinted lights, black chassis, yellow, tan or gray interior, silver hubs, tires, suspension.
1. Silver body, tan interior.
2. Red body, any interior color.
3. Metallic dark red body, tan interior.
4. Orange-brown body.
5. Yellow-green body, tan interior.
6. Light green body.
7. Light metallic blue body.
8. Mustard tan body.

192B ALPINE-RENAULT POLICE CAR 96 mm 1/43 6/1976-1980>#1193

Same basic model as above, with amber dome light, black antenna, gray or tan interior, yellow Gendarmerie decals.
1. Metallic light blue body.
2. Metallic blue body.
3. Purplish-blue body, sometimes metallic.

193 CITROEN GS 95 mm 1/43 2/1972-1980>#1193

Sedan with cast body, matching opening front doors, plastic windows, black chassis and grille, black or tan interior, tinted lights, silver bumpers and hubs, tires, suspension.
1. Rose-gold body.
2. Silver body.
3. White body.
4. Red body.
5. Metallic red body.
6. Yellow-green body.
7. Pale metallic green body.
8. Metallic green body.
9. Metallic greenish-blue body.
10. Light blue body.
11. Metallic light blue body.
12. Metallic sky blue body.
13. Metallic blue body.
14. Metallic purplish-blue body.
15. Metallic blue-gray body.
16. Bronze body.

194 FERRARI 312 PB 84 mm 1.43 5/1972-1980>#1194

Sports-racing car with cast body, matching opening rear hood, silver chassis, plastic windshield, black interior and roll bar, silver motor and hubs, white exhausts, gray mirror, tires, suspension, #51 and other decals.
1. Red body.

195 LIGIER JS 3 85 mm 1/43 3/1972-1976

Sports-racing car with cast body, matching opening rear hood, lower body-chassis, plastic lights, black interior, gray windshield, silver motor, roll bar and hubs, tires, suspension, #24 and other decals.
1. Yellow upper body, green lower body-chassis.

196   RENAULT 17 TS   96 mm   1/43   5/1972-1980>#1196

Coupe with cast body, matching opening doors, plastic windows, black chassis-interior and grille, light gray louvers, silver lights, bumpers and hubs, tires, suspension. Hood +/- bulge.

1. Silver body.
2. Red body.
3. Orange-red body.
4. Yellow body.
5. Metallic green body.
6. Blue body.
7. Metallic blue body.

197   FERRARI 512 M SUNOCO   98 mm   1/43   9/1972-1980>#1197

Sports-racing coupe with cast body, black chassis, plastic windows and lights, white parts, silver airfoil, panels and star hubs, tires, suspension, #11, Sunoco and other decals.

1. Purplish-blue body, Penke or Penske decals.

197B   FERRARI 512 M PIPER   98 mm   1/43   1977-1980>#1198

Same basic model as #197, with #16 and other decals.

1. Red body.

198   PORSCHE 917   96 mm   1/43   12/1972-1980>#1198

Sports-racing coupe with cast body, matching opening rear hood, plastic windows and lights, black chassis-interior, silver motor, cream fan, spoked hubs, tires, suspension, #22, Martini and other decals. Similar to #186 but has rear fins.

1. White body.

199   MARCH 707 CAN-AM   99 mm   1/43   8/1972-1978

Sports-racing car with cast body, plastic windshield, black chassis-interior, green airfoil, silver intakes, roll bar, mirrors and spoked hubs, tires, suspension, #77 and other decals.

1. Red body.

DYNAM SERIES

In 1970 a few Solido models appeared in bubblepacks under the name of Dynam. This name was also cast into their bases, along with a new number. Five of the six models appeared in only one color, the sixth in two:

10 Harvey Aluminum Special: yellow.
11 Fiat Abarth Record Car: red; also white with red cockpit.
12 Alpine Formula 3: blue.
13 NSU Prinz IV: yellow-orange.
14 Abarth 1000 (based on NSU): red.
15 Porsche GT Le Mans: light blue.

THE 200 SERIES

This series is composed almost exclusively of military vehicles. They are fine models and have been very popular among collectors from the start. Many of them were still in production in 1980 and were given new 2200 series numbers, and though Solido stopped making military vehicles for a short time, the 40th anniversary of D-Day in 1984 led to a new life for many of these models in the 6000 series a few years after that.

Having just obtained one of these models in a darker shade of olive than the one I already had, I must add that many of these models could well exist in two shades of olive, not to mention the German "field gray" and the desert-sand tan of specific models.

200 COMBAT CAR M-20 100 mm 1961-1980>#2200
Military vehicle with cast body, chassis, gun mount and hubs, black machine gun and antennas, 6 tires, suspension, French or US decals +/- large number.
  1. Olive body. chassis, gun mount and hubs.

201 UNIC ROCKET TRUCK 177 mm 2/1961-1980>#2201
Truck with cast cab, chassis, rear bed and launcher, unpainted launching mechanism, plastic windows, olive metal or plastic hubs, 10 tires, red or red/white rocket, French rondel decals.
  1. Olive cab, chassis, bed and launcher.

202 PATTON M47 TANK 137 mm 1/1962-1980>#2202 7031
Tank with cast body, matching chassis, turning turret, gun, unpainted tracks, olive plastic hubs, black and gray parts, US or French decals. Also found with plastic turret.
  1. Olive body and chassis, US or French decals.

202S PATTON M47 TANK 137 mm 6/1969-1980>#2214
Same model as above, with Israeli decals.
  1. Tan body.

203 RENAULT 4X4 TOUS TERRAINS TRUCK 96 mm 1962-1980>#2203
Covered truck with cast body, olive cast or plastic hubs, plastic folding windshield, olive front and rear covers, red extinguisher, tires, spare wheel, suspension, #15 decal.
  1. Olive body.

204 105F CANNON 229 mm 1962-1980>#2204
Cannon on cross mount, with cast gun and mount pieces, unpainted elevating wheels, olive plastic parts.
   1. Olive gun and mount.

205 105C CANNON 190 mm 1962-1980>#2205
Cannon on wheels with cast gun, swivel mount and chassis parts, unpainted elevating wheels, olive hubs, tires.
   1. Olive gun, mount and chassis.

206 250/0 HOWITZER 155 mm 1962-1980>#2206
Howitzer on wheels with cast gun, swivel mount and chassis parts, unpainted elevating wheels, olive hubs, tires.
   1. Olive gun, mount and chassis.

207 PT 76 AMPHIBIAN TANK 150 mm 1/1963-1977
Tank with cast body, matching chassis, turning turret, gun and opening hatch, unpainted tracks, olive plastic hubs, red star and number decals.
   1. Olive body.

208 SU 100 TANK 194 mm 3/1964-1980>#2208
Tank with cast body, matching chassis and raising gun, unpainted tracks, gray plastic hubs and parts, red star and number decals.
   1. Olive body.

208S SU 100 TANK 194 mm 1963-1980>#2207
Same model as #208 with Egyptian decals.
   1. Tan body.

209 AMX 30 TANK 170 mm 2/1965-1980>#2209
Tank with cast body, matching chassis, turning turret, gun and opening hatch, unpainted tracks, olive plastic hubs and parts, French decals.
   1. Olive body.

209B AMX 30 TANK 170 mm 10/1976-1980>#2210
Same basic model as #209, with two guns, Egyptian decals.
   1. Tan body.

210 PATTON M47 TANK 137 mm 10/1965-1976
Same basic model as #202, with radio controls. Decals vary.
   1. Olive body.

211 BERLIET T12 TANK TRANSPORTER 337 mm 1/67 1/1967-1980>#2211
Flat semi with cast cab body and chassis, semi chassis and folding ramps, plastic windows, roof light, olive interior, "Transport Exceptionell" roof sign, winch and hubs, tires, suspension, French decals.
   1. Light olive cab and semi.
   2. Dark olive cab and semi.

211B BERLIET TRANSPORTER & AMX 30 TANK 337 mm 1/1967-1980>#2212
Combination of #209 and #211.
   1. Olive tank and transporter.

212 MERCEDES AUTO-UNION JEEP 82 mm 1/43 5/1966-1980>#2213
Jeep with cast body. plastic folding windshield, olive seats, bumper and hubs, tires, suspension.
   1. Olive body.

212C MERCEDES AUTO-UNION JEEP 82 mm 1/43 1968-1978
Same model as #212.
   1. Tan body.

213 JEEP TRAILER 65 mm 1/43 5/1966-1980>#2213

Two-wheel open trailer with cast body, olive plastic hubs, tires. Sold in same box with #212.
   1. Olive body.

213C   JEEP TRAILER   65 mm   1/43   1968-1978
Same model as #213.
   1. Tan body.

214   BERLIET AUROCHS   124 mm   6/1967-1976
Vehicle with cast cab, rear body and chassis, plastic windows, olive hubs, tires, suspension, accessories, OPS decals.
   1. Olive cab, body and chassis.

215   RENAULT 4X4 POLICE TRUCK & MOTORCYCLES   7/1968-1970
Police version of #203, with two motorcycles and accessories.
   1. Dark blue body, gray covers and hubs.

216   AUTO UNION FIRE JEEP AND TRAILER   (1967, not issued)

217   COMBAT CAR   (not issued)

218   PT 76 ROCKET TANK   144 mm   3/1971-1980>#2218
Same basic model as #207 with olive metal and plastic launcher instead of turret, red and white plastic rocket.
   1. Olive body.

219   M41 TANK DESTROYER   134 mm   7/1969-1976
Self-propelled gun with cast body, matching chassis, mount and gun, unpainted elevating wheels and tracks, olive plastic hubs and parts, US decals (serial numbers vary).
   1. Olive body.

220   BSA MOTORCYCLE   (1968, not issued as #220)
Motorcycle used with #215; not sold separately.

221   ALFA ROMEO GTZ POLICE CAR   89 mm   1/43   7/1967-1971
Same basic model as #148 with red dome light, black antenna, French rondel and police decals, plus 2 motorcycles.
   1. Blue body and doors.

222   Pzkw 6 TIGER 1 TANK   166 mm   1/1970-1980>#2222
Tank with cast body, matching chassis, turning turret and gun, unpainted tracks, gray hubs, German decals with #1 or #23.
   1. Dark gray body.

222B   Pzkw 6 TIGER 1 TANK   166 mm   1/1970-1980>#2223
Same model as #222 with camouflage paint.
   1. Tan body with olive camouflage.

223   AMX 13 ANTI-AIRCRAFT TANK   108 mm   3/1970-1976
Tank with cast body, matching chassis and turning turret, unpainted tracks, olive plastic hubs and parts, two black guns, #19 and other decals.
   1. Olive body.

224   XM 706 COMMANDO AMPHIBIAN   114 mm   6/1970-1978
Amphibian with cast body, matching chassis and turning turret, olive engine cover and hatch, tires, suspension, US decals.
   1. Olive body.

224B   XM 706 POLICE VEHICLE   114 mm   6/1970-1978
Police version of #224 with plastic turret, Police decals.
   1. White body and turret, black engine cover.
   2. White body, black turret and engine cover.

225 BTR 40 LANCE-ROCKET CARRIER 114 mm 3/1971-1977

Armored vehicle with cast body, matching chassis and opening hatch, red plastic rockets, olive hubs, tires, suspension, red star and white #416 decals.
1. Olive body.

226 BÜSSING SdKfz 232 SCOUT CAR 114 mm 5/1975-1980>#2226

Armored car with cast body, matching chassis and turning turret, black plastic antenna, gray hubs, tires, suspension, German decals. **7030**
1. Dark gray body.

227 AMX 13 V.C.I. 112 mm 4/1971-1976

Armored vehicle with cast body, matching chassis, turret and opening hatches, unpainted tracks, olive plastic hubs, red cross decals. Turret does not turn.
1. Olive body.
2. Khaki body.

227B AMX 13 V.T.T. 112 mm 9/1972-1980>#2227

Same model as #227 plus turning turret with gun, "Franche-Comté" or "Hombourg" decals.
1. Olive body.

228 JAGDPANTHER TANK 196 mm 12/1971-1980>#2228

Tank with cast body, matching chassis and gun, unpainted tracks, gray plastic hubs, black exhausts, decals with #123 or #474.
1. Dark gray body.

228B CAMOUFLAGED JAGDPANTHER 196 mm 12/1971-1980

Same model as #228, with camouflage paint.
1. Tan body with maroon camouflage.

228N WHITE JAGDPANTHER 196 mm (1974, not issued)

229M JAGDPANTHER TANK 196 mm 12/1971-1977

Same model as #228 plus radio control, "Hombourg" decals.
1. Tan body with maroon camouflage.

230 AMX 13 TANK WITH 90 GUN 135 mm 1/1972-1976

Tank with cast body, matching chassis and turning turret, unpainted tracks, olive plastic hubs, black parts, "Hombourg" decals.
1. Olive body.

231 SHERMAN M4 TANK 124 mm 3/1972-1980>#2231

Tank with cast body, matching chassis, turning turret and gun, unpainted tracks, olive plastic hubs, black gun, US decals with number 12, 29 or 670.
1. Olive body.

232 M-10 TANK DESTROYER 123 mm 6/1972-1980>#2232

Armored vehicle with cast body, matching chassis, turning turret and gun, unpainted tracks, olive plastic hubs, black gun, US (with #179) or French (with "Alsace") decals.
1. Olive body.

233 RENAULT R35 TANK 95 mm 2/1973-1980>#2233

Tank with cast body, matching chassis, turning turret and gun, unpainted tracks, olive plastic hubs and parts, French roundel decals.
1. Two-tone gray camouflage body and turret.
2. Olive and brown camouflage body and turret.

234 SOMUA S35 TANK 107 mm 3/1973-1980>2234

Tank with cast body, matching chassism turning turret and gun, unpainted tracks, olive plastic hubs, black parts, French roundel decals.
1. Olive and brown camouflage body and turret.

235 SIMCA-UNIC 4X4 TRUCK & 105 CANNON 244 mm 6/1973-1979

Covered truck with cast cab and rear body-chassis, pulling cannon with cast gun and chassis, plastic windshield, olive rear cover, driveshafts and hubs, tires, suspension.

1. Olive truck and cannon.

236 PANTHER G TANK   175 mm   12/1973-1980>#2236

Tank with cast body, matching chassis, turning turret and gun, unpainted tracks, gray plastic hubs, black parts, German decals with #421 or #137.

1. Dark gray body, #421.
2. Tan and maroon camouflage body.

237 PANZER IV TANK   142 mm   5/1974-1980>#2237

Tank with cast body, matching chassis, turning turret and gun, unpainted tracks, gray plastic hubs, German decals with #123, #240 or #474.

1. Dark gray body.

238   AMX PLUTON MISSILE LAUNCHER   ___ mm   9/1977-1980>2238

Missile launcher on tracks, with missile, "Douai" decals.

1. Olive body.

239 105 CANNON   141 mm   6/1973-1979

Cannon of #235; listed but not issued separately. See #235.

240   PANHARD AML H90 ARMORED CAR   108 mm   10/1973-1980>#2240

Armored vehicle with cast body, matching turning turret and gun, matching plastic chassis, front plate, tank and hubs, tires, "Picardie Belfort Flandre" decals around turret.

1. Olive body.

240S PANHARD AML H90 ARMORED CAR   108 mm   3/1974-1980

Same model as #240, but with black #891 decals.

1. Tan body.

241   HANOMAG SdKfz 251/1 HALFTRACK   120 mm   2/1974-1980>#2241

Halftrack with cast body, matching chassis, unpainted tracks, gray plastic hubs, black machine guns, front tires, German decals.

1. Dark gray body.

242   DODGE 6X6 ARMY TRUCK   120 mm   3/1975-1980>#2242

Truck with cast chassis-cab and rear body, olive plastic cover, grille and hubs, folding windshield, tires, suspension, US decals.

1. Olive cab and body.

243 LEOPARD KPZ TANK   180 mm   1/1975-1980>#2243

Tank with cast body and turning turret, unpainted tracks, German decals with #621 or #137.

1. Olive body.

244 M3 HALFTRACK   127 mm   5/1976-1980>#2244

Halftrack with cast body, matching chassis, olive plastic cover and hubs, black tracks, front tires, US decals.

1. Olive body.

245   KAISER-JEEP 6X6 ARMY TRUCK   140 mm   11/1975-1980>#2245

Truck with cast body parts, matching chassis, plastic windows, olive cover and hubs, tires, suspension. +/- US decals.

1. Olive body.

246 T34/85 TANK (announced as #234, then #246, not issued)

247 BERLIET ALVIS 125 mm 5/1975-1980>#2247
Six-wheel vehicle with cast body parts, matching chassis and opening hatches, plastic windows, olive hubs, black antenna, tires, suspension, +/- OPS decals.
1. Olive body.

248 M41 TANK DESTROYER ___ mm 1976-1980>#2248
Gun vehicle with cast body and turning turret, unpainted tracks, US decals.
1. Olive body.

249 AMX 13 TWO-GUN TANK 110 mm 6/1975-1980
Tank with cast body, matching chassis and turning turret, unpainted tracks, black plastic guns, olive body parts and hubs, French decals with #19.
1. Olive body.

250 AMX 13 TANK WITH 90 GUN 134 mm 5/1975-1980>#2250
Tank with cast body, matching chassis and turning turret, unpainted tracks, black plastic parts, olive hubs, French decals which may include "Franche-Comté" or "Hombourg".
1. Olive body.

251 SAVIEM VAB 6X6 AMPHIBIAN 120 mm 11/1976-1980>#2251
Amphibian with cast body parts, six wheels.
1. Olive body.

252 M7 B1 ARMORED VEHICLE 118 mm 12/1976-1980>#2252
Gun vehicle with cast body, matching chassis and parts, unpainted tracks, black plastic machine gun, olive hubs, figure, US decals with #27, #45 or #52.
1. Olive body.

253 GENERAL LEE TANK 116 mm 7/1978-1980>#2253
Tank with cast body, matching turning turret and guns, unpainted tracks, olive plastic chassis and hubs, US decals.
1. Olive body.

254 AMX 10P ARMORED VEHICLE 115 mm 3/1978-1980>#2254
Armored vehicle with cast body, unpainted tracks, French decals.
1. Olive body.

255 BERLIET CAMIVA FOAM TRUCK 140 mm 5/1979-1980>#2255
Tank truck with cast body, matching chassis, plastic windows, black nozzle and ladder, olive hubs, tires, suspension, Armee de l'Air decals.
1. Olive body.

256 WILLYS JEEP AND TRAILER 140 mm 5/1979-1980>#2256
Jeep with cast body, two-wheel trailer with cast body and chassis (=#213), plastic folding windshield, olive top, tank and chassis, tires, spare wheel, +/- US decals.
1. Olive Jeep and trailer.

257 SAVIEM SR TANK TRUCK 215 mm 1/60 10/1979-1980>#2257
Tanker semi with cast cab, both chassis and catwalk, plastic windows, black ladder, olive tank and hubs, tires, spare wheel, yellow label with black F-34.
1. Olive cab, chassis and catwalk.

258 BERLIET GAZELLE TRUCK   (announced 1978, not issued)

259 CITROEN C 35 ARMY AMBULANCE 98 mm 1/50 2/1978-1980>#2259
Ambulance van with cast body, matching chassis and opening rear doors, plastic windows, blue dome light, olive interior and hubs, tires, red cross decals.
1. Olive body and roof.
2. Olive body, white roof.

260 GAZELLE HELICOPTER   (not released; see 3815)

261 AMX 10 RC TANK   (announced 1979, not released)

262 RICHIER ARMY CRANE TRUCK 140 mm 5/1979-1980>#2262
Truck with cast body, matching chassis, swiveling mount and main boom, unpainted telescopic boom, tinted plastic windows, black hook and parts, olive hubs, tires, spare wheel.
1. Olive body.

THE 300 SERIES

This series of commercial vehicles began with several very impressive semi-trailers and other heavy trucks. It grew to include a wide variety of models, and when the new numbering system took effect, it was seen that they did not all belong in the same class in terms of size and cost, and they were eventually divided among several 2000 and 3000 series.

300   BERLIET TBO 200 CV TITAN   293 mm   1/1960-1972
Semi with cast cab, chassis, semi and matching wheel mount, plastic windows, white box with gray winch and spare tires, gray exhaust pipe, stakes and hubs, tires, suspension, silver grille, load of red and white oil well parts.
  1. Tan cab and semi.

300B   BERLIET TBO 300 CV TITAN   293 mm   6/1961-1972
Same basic model as #300, but cab has black fenders.
  1. Tan cab and semi.

301   UNIC SAHARA TITAN TANK TRUCK   295 mm   1961-1972
Semi with cast cab, chassis, semi and matching wheel mount, plastic windows, white box with gray winch and spare tires, gray hubs, tires, suspension, load of four tanks with "Eau Potable", "Farine" or "Ciment" lettering. Tanks all white or white with red tops. Green cylindrical "Eau Potable" tanks also exist.
  1. Red cab and semi.
  2. Yellow cab and semi.
  3. Light green cab and semi.
  4. Blue cab and semi.
  5. Tan cab and semi.
  6. Light gray cab and semi.
  7. Dark gray cab and semi.

302 WILLEME HORIZON TITAN OPEN TRUCK   268 mm   6/1961-1972
Open semi with cast cab, chassis, semi bed, sides and wheel mount matching semi, plastic windows, hubs, tires, suspension, silver grille and lights.
  1. Red cab and sides, gray chassis and bed.
  2. Yellow cab and semi.

3. Green cab and semi.
4. Blue cab and semi.
5. Tan cab and semi.

303 BERLIET QUARRY DUMP TRUCK 160 mm 1/43 1962-1978

Dumper with cast cab, tipper, gray or silver chassis, unpainted tipping mechanism, plastic windows, black or gray exhaust, gray hubs, tires, suspension, silver grille and lights.
1. Red cab and tipper.
2. Brick red cab and tipper.
3. Yellow cab and tipper.
4. Green cab and tipper.
5. Olive cab and tipper.
6. Tan cab and tipper.

304 BERNARD T12 REFRIGERATOR TRUCK 176 mm 1/43 1963-1972

Truck with cast cab, lower rear body, second-color box and opening door, gray chassis, plastic windows, gray hubs, tires, suspension, silver grille and lights, red-white-blue "La vache qui rit" and "Fromageries Bel" decals (2 similar types).
1. Red cab and lower body, bluish-white box.
2. Red cab and lower body, cream box.
3. Light blue cab and lower body, white box.

305 BERLIET T12 SR LOW LOADER 336 mm 1/67 5/1967-1977

Low loader semi with cast cab, chassis, semi, matching ramps and winch bracket, unpainted wheel mounts, plastic windows, roof light, gray interior, winch and hubs, tires, two spare wheels, load of gray and white prefab building parts, "Algeco" logo.
1. Dark yellow cab and semi.
2. Light yellow cab and semi.

305B BERLIET T12 SR LOW LOADER 336 mm 1/67 1977-1980>#3305

Same model as #305, with pipes bearing "GDF" decals, red Renault 4 van, "STE" decals on cab and R4, policeman figure.
1. Red cab and semi, silver cab chassis.

306 BERLIET STRADAIR DUMP TRUCK 156 mm 1/43 3-1967-1973

Dumper with cast cab, chassis, tipper, tailgate, cab floor, unpainted tipping mechanism, plastic windows, black interior, gray or red hubs, tires, suspension, spare wheel.
1. Silver cab, chassis and tipper, red floor.
2. Silver, red and yellow.
3. Silver, blue and yellow.
4. Silver, blue and red.
5. Red cab and chassis, light yellow tipper and floor.
6. Turquoise and light gray.

307 BERLIET STRADAIR COVERED TRUCK 155 mm 1/43 10/1968-1973

Covered truck with cast cab, chassis, rear body with matching opening doors, red cab floor, plastic windows, charcoal or green cover, gray hubs, tires, suspension, grille and logo decals.
1. Silver cab and chassis, yellow rear, "Trigano" logo.
2. Colors as above, "Les Déménageurs Rapides" logo.

308 WILLEME ELF TANK TRUCK 142 mm 4/1968-1972

Tanker with cast cab, chassis, matching hubs, plastic windows, blue rear bed, white tank, red caps and valve, tires, suspension, silver grille and lights, red and black "elf" logo decals.
1. Turquoise blue cab and rear.
2. Dark blue cab and rear.
3. Light blue cab and rear.

316 SAVIEM SM 300 OPEN SEMI WITH CRANE 275 mm 7/1972-1979

Semi with cast cab, chassis, semi bed, sides, light gray wheel mount, unpainted crane parts, plastic windows, tan interior, cream or green load (4 sections), black steering wheel and crane hook, gray hubs, tires, suspension.

    1. Red cab and sides, silver gray chassis and bed, cream load.
    2. Blue cab and sides. red chassis and bed, green load.

317 BERLIET TR 300 SR TRANSPORTER 275 mm 5/1972-1979

Semi with cast cab, semi, silver chassis and wheel mount, plastic windows, tinted lights, tan interior, silver parts, green hubs, tires, suspension, spare wheel, silver grille and bumper, load of five off-white tanks, one with black "Yoplait" logo, one with flower design.

    1. White cab, green semi.

318 SAVIEM SM 300 ELF TANKER 150 mm 5/1972-1977

Tanker with cast tiltcab, silver chassis, plastic windows, tinted lights, tan interior, blue rear bed, white tank, red caps and valve, blue hubs, tires, suspension, silver bumper, grille and red-black "elf" logo decals.

    1. Blue cab.

319 SAVIEM SM 300 ESSO TANKER 150 mm 5/1972-1977

Same basic model as #318, with gray rear bed and hubs, red-white-blue "Esso" logo.

    1. White cab and tank.

320 SAVIEM SM 300 SHELL TANKER 150 mm 5/1972-1977

Same basic model as #318, with red rear bed and hubs, red-yellow "Shell" logo.

    1. Yellow cab and tank.

321 SAVIEM CAR CARRIER & TRAILER 424 mm 12/1974-1980>#3321

Carrier and trailer with cast cab, yellow upper and silver lower decks, trailer with yellow upper and silver lower decks and ramp, unpainted axle mounts and hubs, plastic windows, tan interior, gray hubs, tires, suspension, grille and "Causse Walon" decals.

    1. Silver cab.
    2. Red cab.
    3. Yellow cab.
    4. Blue cab.
    5. Metallic blue cab.
    6. Blue and silver.
    7. Blue and yellow.

330 CITROEN C 35 CIRCUS VAN 112 mm 3/1979-1980>6330

Van with cast body, roof and black chassis, plastic windows, red interior, balcony, loudspeaker and grille, figure, wheels, "Amar Service Publicité" decals.

    1. White body, red roof.

331 MERCEDES CIRCUS TRUCK 123 mm 1/1979-1980>#6331

Covered truck with cast cab, chassis and rear body with tailgate, plastic windows, red cover, black interior-grille, inner chassis and front fenders, wheels, "Amar" labels.

    1. Red cab and chassis, white rear body.

332 SAVIEM FLAT SEMI WITH CAGES 223 mm 3/1979-1980>#6332

Flat semi with cast cab, chassis and semi, blue plastic windows, white semi chassis and cages, black interior-grille, lion and tiger figures, wheels, "Amar" labels.

    1. Red cab and semi, white chassis.

333 DAF CIRCUS BOX OFFICE SEMI 223 mm 3/1979-1980>#6333
Semi with cast cab, chassis and semi chassis, tinted plastic windows, white semi body, roof sign, opening door and inner chassis, black cab interior-grille-fenders and inner chassis, wheels, "Amar" and "Caisse" labels.
1. Red cab and semi chassis, white cab chassis.

334 RICHIER CIRCUS CRANE TRUCK 139 mm 1/1979-1980>#6334
Crane truck with cast body, white chassis and boom, unpainted mount and telescopic boom, blue plastic windows, white hook and parts, wheels, spare wheel, "Amar" decals.
1. Red body.

335 DAF ANIMAL TRANSPORTER SEMI 228 mm 6/1979-1980>#6335
Semi with cast cab, chassis, and semi chassis, tinted plastic windows, white semi body, ramp and inner chassis, black cab interior-grille-fenders, wheels, "Amar" labels.
1. Red cab and semi chassis, white cab chassis.

336 DAF STAKESIDE CIRCUS SEMI 228 mm 7/1979-1980>#6336
Semi with cast cab, chassis and semi bed, tinted plastic windows, white semi stake body and inner chassis, black cab interior-grille-fenders and inner chassis, wheels, "Amar" labels.
1. Red cab and semi, white cab chassis.

337 DAF CIRCUS CARAVAN SEMI 228 mm 7/1979-1980>#6337
Semi with cast cab, chassis and semi chassis, tinted plastic windows, white semi body, opening door, and inner chassis, black cab interior-grille-fenders and inner chassis, wheels.
1. Red cab and semi, white cab chassis.

338 STAKESIDE CIRCUS TRAILER 194 mm 6/1979-1980>#6338
Trailer with cast bed, white plastic stake body and inner chassis, red hitch, wheels, "Amar" labels.
1. Red bed.

350 BERLIET CAMIVA FIRE ENGINE 123 mm 11/1972-1980>#3350>#2107
Fire truck with cast body, blue plastic windows and dome light, red chassis and rear panel, red and white hose reels, silver ladder, wheels, "Ville de Paris" decals. Early type has battery space and working siren.
1. Red body.

351 BERLIET AIRPORT FIRE TRUCK 140 mm 12/1972-1980>#3351>#3107
Tanker with cast body, black chassis, blue plastic windows, silver nozzle, ladder and catwalks, wheels, yellow roof panel, "Aeroport de Paris" or "Nice-Côte d'Azur" decals.
1. Red body.

352 BERLIET LADDER TRUCK 141/163 mm 3/1973-1980>#3352>#3109
Ladder truck with cast body, mount, silver 2-piece ladder, black chassis, blue plastic windows, black and silver rear panels, wheels, "Ville de Paris" decals.
1. Red body and mount.

353 RICHIER CRANE TRUCK 139 mm 1/1973-1980>#3353>#3102
Crane with cast body, matching boom, unpainted mount and telescopic boom, black chassis, tinted plastic windows, gray hook, wheels, spare wheel, "Nordest" or "Richier" labels.
1. Red body.
2. Yellow body.

354 BERLIET FOREST FIRE TRUCK & TRAILER 145 mm 12/1973-1980>#3354

Fire truck with cast body and black chassis, trailer with cast body and matching chassis, blue plastic windows and dome light, silver hose reel, yellow roof panel, silver or red-hub wheels on truck, hubs and tires on trailer, spare wheel, "Service Departemental d'Incendie" decals.
1. Red bodies.

355 PEUGEOT J7 MINIBUS 92 mm 1/50 5/1976-1980

Minibus with cast body, opening doors, white roof and black chassis, plastic windows, tan interior, wheels, +/- red-blue "Air Inter" decals.
1. Purplish-blue body and doors.
2. Light blue body and doors.

355B PEUGEOT J7 SCHOOL BUS 92 mm 1/50 5/1976-1980

Same model as #355, +/- yellow-on-black "Transport d'Enfants" decals.
1. Light cream body, white roof.
2. Pale green body, white roof.

356 VOLVO-BM DUMP TRUCK 160 mm 1/1974-1980>#3356>#3103

Dumper with cast cab parts, tipper and chassis, unpainted joint and axle mounts, tinted plastic windows, wheels. blue "Volvo-BM" decals.
1. Yellow cab and tipper.

357 UNIC SAHARA DUMP TRUCK 158 mm 4/1974-1980

Quarry dumper with cast cab, chassis, tipper, unpainted axle mounts, black bumper-cab floor, plastic windows, wheels.
1. Dark yellow cab, chassis and tipper.

358 MERCEDES OVERHEAD SERVICE TRUCK 149/175 mm 7/1974-1980 >#3358>#3105

Truck with cast body, mount, lower and upper arms, basket, black, silver or red chassis-deck panel-bumper, blue or amber plastic windows (with dome light and black antenna in some versions), wheels, "Reportage Télévise" or "Travaux Électriques" decals, grille either cast-in or decal.
1. Red body, yellow mount, arms and basket.
2. Red-orange body, otherwise as type 1.
3. Orange body, otherwise as type 1.
4. Light blue body, silver arms, yellow mount and basket.
5. Blue body, otherwise as type 4.
6. Metallic blue body, red chassis, yellow mount and arms, orange two-section basket with "Egi" decals.

359 SIMCA-UNIC SNOWPLOW TRUCK 142 mm 3/1974-1980>#3359>#2108

Covered truck with cast body, silver plow, unpainted drive train, plastic windshield, orange or yellow cover, black parts, wheels with gray tire chains, stripe labels.
1. Red body, yellow cover.
2. Yellow body, orange cover.

360 GUINARD FIRE PUMP TRAILER 54 mm (12/1973-1981)

Trailer with cast body and chassis, silver hubs, tires. Part of #354 but not sold separately.
1. Red body and chassis.

361 MERCEDES LADDER TRUCK 150/163 mm 10/1975-1980>#3361>#3101

Ladder truck with cast body (= #358), matching mount, black or silver chassis-deck panel-bumper, unpainted ladders (= #352), blue or amber plastic windows +/- dome lights and black antenna, wheels, cast or decal grille, red-on-white "Service Departemental" decals.
1. Red body.

362   HOTCHKISS FIRE ENGINE   111 mm   3/ 19751980>#3362>#2000> #2100

Open-cab fire truck with cast body, matching pump, black chassis, plastic windshield, red and white hose reel, silver panels and front bumper, black interior and reel, wheels, "Service Departemental d'Incendie" decals.
1. Red body.

363 MAGIRUS FRUEHAUF COVERED SEMI   250 mm   9/1975-1980>#3363

Covered semi with cast cab, silver and yellow chassis, yellow semi and opening doors,, silver chassis, plastic windows, dark blue cover, , green cable, wheels, two spares, "Solido Export" or "Rentco" logo labels, or no labels. Later versions have exhaust stack, black or maroon cover.
1. Metallic red cab, yellow semi, blue cover, either or no logo.
2. Yellow cab and semi, blue cover, "Rentco" logo.
3. Yellow cab and semi, black or maroon cover, no logo.

363B DAF F2800 COVERED SEMI 259 mm 6/1978-1980>#3501

Covered semi with cast cab, black cab floor-fenders, silver chassis, semi, opening doors and chassis, blue plastic windows, silver horns, brown cover, wheels, two spares, white "Onatra" logo labels.
1. Orange cab.
2. Yellow cab.

364   MERCEDES 2624 BUCKET TRUCK   136 mm   11/1975-1980>#3364

Truck with cast cab, black bumper-fenders, silver or yellow chassis, silver rear body and arms, yellow bucket, plastic windows, black parts, wheels, +/- "Bennes Marrel" decals.
1. Metallic dark red cab, silver chassis.
2. Dark blue cab, yellow chassis.

365 INTERNATIONAL POWER SHOVEL   __ mm   3/1976-1980>#3365

Excavator with IH and Yombo logo.
1. Yellow body.

366   SAVIEM SG4 WRECKER   105 mm   1/50   5/1977-1980>#3300>#2006

Wrecker with cast cab, body, red or unpainted frame, unpainted boom, blue or amber plastic windows and roof lights, white towhooks, black chassis, wheels, various decals.
1. White cab, red body, "SOS Dépannage" logo.
2. White cab, light blue body, "SOS Dépannage" logo.
3. Light blue cab and body, "SOS Dépannage" logo.
4. Dark blue cab and body, "SOS Dépannage" logo.
5. Red cab and body, "Service Departemental d'Incendie" logo.
6. Dark blue cab and body, "Police" logo.

367   VOLVO-BM SHOVEL LOADER   129 mm   1/55   2/1977-1980>#3367 >#3104

Loader with cast cab, engine cover, front and rear chassis, arms and shovel, unpainted exhaust pipe and axle mounts, tinted plastic windows, black parts, wheels, +/- "Volvo-BM" decals.
1. Yellow body parts and shovel.
2. Orange and silver.

368   CITROEN C35 FIRE AMBULANCE   96 mm   1/50 3/1977-1980>#3368 >#2116

Van with cast body, opening rear doors, roof, black chassis, plastic windows, blue dome light, white interior and grille, wheels, circular "Service Départemental d'Incendie" or one-line "Service Départemental" decals.
1. Red body and doors, white roof.

369 DAF F2800 TANKER SEMI 217 mm 1/60 3/1977-1980>#3369 >#3500

Tanker semi with cast cab, inner and outer chassis, semi body, chassis and parts, red catwalk, plastic windows, off-white tank, silver horns, black ladder, wheels, spare, logo decals.
1. Orange cab and semi, black and silver chassis, Shell logo.
2. Yellow cab and semi, silver chassis, Shell logo.
3. Yellow cab and semi, silver chassis, Onatra logo.

370 SAVIEM H875 SEMI 218 mm 11/1977-1980>#3370>#3502

Box semi with cast cab, inner & outer chassis, and semi bed, blue plastic windows, white box, roof, opening doors and semi chassis, black grille, wheels, "Renault" labels (two types).
1. Yellow cab, chassis and semi.
2. Yellow cab and semi, black chassis.

371 CITROEN AMBULANCE & LIFEBOAT TRAILER 192 mm 1/50 7/1977-1980>#3371>#3100

Same basic model as #368, with white "Secours aux noyes" decals, pulling cast trailer with brown and gray plastic boat, white hubs, tires.
Red ambulance budy and trailer, white roof.

372 PEUGEOT J7 POLICE BUS 91 mm 1/50 3/1977-1980>#3372>#2003

Minibus with cast body, opening doors, roof, black chassis, plastic windows, gray interior and siren, blue dome light, wheels, black-on-white "Police" decals.
1. Blue body and doors, white roof.

373 MERCEDES 1217 LIVESTOCK TRUCK 120 mm 1/55 4/1978-1980>#3373

Truck with cast cab, rear body, matching ramp, silver outer and inner chassis, blue plastic windows, white top, wheels.
1. Green cab and body.

374 UNIC IVECO DUMP TRUCK 135 mm 1/60 10/1978-1980>#3374 >#2009>#2109

Dumper with cast cab, tipper, matching tailgate, black chassis, clear or tinted plastic windows, black parts, black or silver and black grille, +/- silver horns and interior, wheels.
1. Red cab, yellow tipper, silver horns and interior.
2. Dark orange cab and tipper, no horns or interior.

375 BERLIET GAK FIRE ENGINE 116 mm 1/55 11/1978-1980>#3375 >#2008>#2106

Fire truck with cast body, blue plastic windows and dome light, white grille and roof panel, red and white hose reels, silver horns, black chassis, wheels, "Ville de Paris" decals.
1. Red body.

376 MERCEDES BULK CARRIER SEMI 211 mm 1/60 12/1978-1980> #3376

Semi with cast cab, semi chassis, plastic windows, black cab chassis, white tank, catwalk and ladder, wheels, grille and "Ciment en vrac" labels.
1. Red cab, silver semi chassis.

377 MERCEDES SNORKEL FIRE TRUCK (announced 1978, not issued)

378 MERCEDES EXCAVATOR TRUCK 173 mm 2/1978-1980>#3378

Excavator truck with cast body, matching cab, engine cover, arms and shovel, charcoal chassis-deck panel, plastic windows and roof lights, black parts, wheels, grille label.
1. Orange body and matching parts.

379 MERCEDES 1217 GARBAGE TRUCK 140 mm 2/1979-1980>#3379

Truck with cast cab, rear body and matching hinged rear, plastic windows, black inner and outer chassis and parts, wheels.

   1. Brown cab, tan rear body.

380 PEUGEOT J7 FIRE AMBULANCE 93 mm 1/50 1/1978-1980>#3380 >#2115

Minibus with cast body, matching opening doors, roof, blue plastic windows and dome light, gray or tan siren and interior, wheels, white "Ambulance Departmentale" decals.

   1. Red body and roof.

   2. Red body, white roof.

381A GAZELLE HELICOPTER 171 mm 1/55 4/1979-1980>#3810

Copter with cast body, unpainted interior, tinted plastic windows, black rotors and parts, figure, decals.

   1. White body, "europ assistance" decals.

   2. Blue-black body, Italian roundel and "Carabinieri" decals.

381B GAZELLE POLICE HELICOPTER 171 mm 1/55 4/1979-1980> #3811

Same model as #381A, with different colors and decals.

   1. Blue body, white "Gendarmerie" decals.

   2. Dull blue body, Italian roundel and "Polizia" decals.

382 MERCEDES REFRIGERATOR TRUCK (announced 1979, not issued)

383 MERCEDES MOVING VAN (announced 1979, not issued)

384 MERCEDES COVERED TRUCK 121 mm 1/55 1/1979-1980>#3384

Covered truck with cast cab, rear body, matching tailgate, red chassis, blue plastic windows, pale gray cover, black inner chassis and grille-fenders, wheels, "Transports toutes distances" labels.

   1. Silver cab, dark blue body.

385 SAVIEM HORSE VAN SEMI 229 mm 9/1979-1980>#3385

Semi with cast cab, chassis, semi chassis, plastic windows, brown semi body, ramp and inner chassis, black inner cab chassis-grille, wheels, yellow "STCC" labels.

   1. Silver cab and both chassis.

386 MERCEDES PROPANE TANKER 127 mm 1/55 9/1979-1980>#3386

Truck with cast cab and 2-piece chassis. blue plastic windows, white tank, black grille-fenders and inner chassis, wheels, "Propane Butagaz" labels.

   1. Metallic blue cab and chassis.

388 SAVIEM STAKESIDE SEMI 229 mm 4/1979-1980>#3388

Semi with cast cab, chassis and semi bed, blue plastic windows, white stake body and semi chassis, black inner cab chassis, wheels, brown "Europeenne de Transport" labels, log load.

   1. Lime green cab, chassis and semi.

389 STAKESIDE TRAILER 195 mm 4/1979-1980>#1980

Four-wheel trailer based on #388 semi, with cast bed, white stake body and chassis, black hitch-fenders, wheels, brown "Europeenne du Transport" labels.

   1. Lime green bed.

390 DAF F2800 SEMI (announced 1979, not issued)

391 DODGE 6X6 FIRE TRUCK & TRAILER 180 mm 11/1978-

1980>#3391

Open truck and trailer with cast bodies and chassis, plastic windshield, black grille, wheels, spare wheel.

    1. Red bodies and chassis.

THE 500 SERIES

There would have been a 400 series, composed of two buses, if the new numbers had not been introduced when they were, but those two buses did not appear until just after new numbers were given to them. But there was a 500 series of farm machinery, which had been introduced in 1977.

510 RENAULT FARM TRACTOR 109 mm 1977-1980>#3510
Tractor with cast body parts, unpainted steering column, tinted plastic windows, white cab, black parts and tires, cream hubs.
  1. Red body.
  2. Orange body.

511 TIPPING FARM TRAILER 180 mm 1977-1980>#3511
Two-wheel trailer with cast body and chassis, white plastic tailgate, white and black parts, cream hubs, tires.
  1. Red body and chassis.

512 RENAULT TRACTOR & TIPPING TRAILER 282 mm 1977-1980>#3512
Combination of #510 and #511.
  1. Red bodies and chassis.

513 TANK TRAILER 140 mm 1977-1980>#3513
Two-wheel trailer with cast chassis, white plastic tank, red caps and valve, cream hubs, tires.
  1. Orange chassis.

514 RENAULT TRACTOR & TANK TRAILER 243 mm 1977-1980>#3514
Combination of #510 and #513.
  1. Orange body and chassis.

515 SILAGE TRAILER 96 mm 1977-1980>#3515
Two-wheel trailer with cast body, white plastic parts, cream hubs, tires.
  1. Orange body.

516 SPRAYER TRAILER 104 mm 1977-1980>#3516
Two-wheel trailer with cast chassis, red plastic tank, green sprayers, cream hubs, tires.
  1. Green chassis.

THE 600 SERIES

The 600 series began with three 1/32 scale cars, issued singly and with trailers in 1970. After the large-scale models had been dropped, the concept of car and trailer remained, augmented by sets of several models, to keep the series alive.

600 PEUGEOT 504 127 mm 1/32 3/1970-1973
Sedan with cast body, matching opening trunk, black chassis, plastic windows and lights, red or tan interior, silver grille and hubs, tires, suspension. Can have tow hook (see #610).
1. Metallic pale gold body.
2. Metallic red body.
3. Metallic light blue body.

601 OPEL COMMODORE 132 mm 1/32 3/1970-1973
Hardtop with cast body, matching opening trunk, black chassis, plastic windows and lights, red or tan interior, silver grille and hubs, tires, suspension. Can have tow hook (see #611).
1. Metallic light gold body.
2. Metallic green-gold body.
3. Metallic light gray body.

602 RENAULT 16 124 mm 1/32 3/1970-1973
Hatchback sedan with cast body, matching opening hatch, black chassis, plastic windows and lights, red or tan interior, silver grille, bumpers and hubs, tires, suspension. Can have tow hook (see #612).
1. Metallic red body.
2. Turquoise blue body.

610 PEUGEOT 504 & STAR CARAVAN ___ mm 1/32 1970-1980
Combination of #600 and caravan trailer.
1. Metallic red car, white trailer.
2. Metallic blue car, white trailer.

611 OPEL COMMODORE & BOAT TRAILER 247 mm 1/32 1970-1973
Combination of #601 and two-wheel trailer with cast body, blue or yellow plastic blocks, silver hubs, tires, carrying boat with plastic windshield, red and white parts.
1. Metallic light gold car, red or light blue trailer.
2. Metallic green gold car, red or light blue trailer.
3. Metallic light gray car, red or light blue trailer.

612 RENAULT 16 & ANIMAL TRAILER 226 mm 1/32 1970-1978
Combination of #602 and #197 trailer with cast body, red plastic stakes, silver hubs, tires, figures.
1. Metallic red car, yellow trailer.
2. Metallic blue car, yellow trailer.

613 RENAULT 12 & POLICE MOTORCYCLE 101 mm 1/43 5/1975-1980>#7613
Police version of #22 with amber dome light, white "Gendarmerie" decals, plus plastic motorcycle with driver, policeman figure, white box, green base.
1. Dark blue car, black and silver cycle.

614 CITROEN CX & DATCHA TRAILER ___ mm 1/43 11/1975-1980
>#7614
#29 car with Sterckeman Datcha 530 trailer (see #621 below).
1. Metallic red car, white trailer.

615 FIAT X1/9 & MOTORCYCLE TRAILER 156 mm 1/43 6/1975-1980>#7615
#33 car with plastic trailer, same hubs and tires as car, cast motorcycle with silver and gray plastic parts, racing number.
1. Metallic dark red car, blue trailer, yellow cycle.

616 PEUGEOT 504 & HORSE TRAILER ___ mm 1/43 4/1977-1980 >#7616
#23c wagon with trailer and horse figure.
1. Silver car, green and tan trailer with silver chassis.

617 WINTER SPORTS SET 1/43 12/1977-1980>#7617
Simca-Unic #359 truck, Lancia Beta #52 with ski rack, and skier figure.

1. Yellow and silver truck, metallic light green car.

618 RACING TEAM SET 226 mm 1/43 3/1978-1980>#7618
Renault 12 #22 wagon with cast trailer and #58 Renault 5TL with racing decals.
1. Light blue R12, silver trailer, yellow R5.

619 RENAULT 30 & BOAT TRAILER 196 mm 1/43 4/1978-1980 >#7619
Renault #30 with tan plastic surfboard rack, pulling cast trailer with brown and gray plastic boat (as in #371).
1. Green car, silver trailer.

620 WRECKER AND LANCIA BETA 205 mm 1/43 4/1978-1980>#7620
Saviem #366 wrecker with "SOS Dépannage" decals, Lancia Beta #52, and mechanic figure.
1. Orange wrecker, green car.

621 CIRCUS CARAVAN SET 248 mm 1/43 7/1979-1980>#6621
Land-Rover #66 and Sterckeman trailer as in #610, both with "Amar" circus logo labels.
1. White and red Land-Rover and trailer.

650 FARM SET 11/1978-1980>#3650
Boxed set of farm vehicles and accessories.

660 CIRCUS SET 11/1978-1980>#6660
Boxed set of circus vehicles and accessories.

THE 10 SERIES

When the 100 series ran out of numbers, Solido introduced a new two-digit series of cars in 1973. The first number used was 10, though a few single-digit numbers were used at the very end. Most of the 10 series models became 1000 series models in 1980, and some lived on to appear in the 1300 series a few years later. The details of models 1 to 7 will be found in the 1000 and 1300 listings, where they spent most, if not all, of their lives.

1 RENAULT 4 PTT VAN   89 mm   1/43   1980>#1001

2 RENAULT 4 FIRE VAN   89 mm   1/43   1980>#1002

5 CITROEN VISA   85 mm   1/43   1980>#1302

6 FIAT RITMO   89 mm   1.43   1980>#1303

7 CITROEN 2CV   88 mm   1980>#1301

10   RENAULT 5 TL   78 mm   1/43   6/1972-1980>#1010
Two-door hatchback with cast body, matching opening doors, plastic windows and lights, gray-brown chassis, red, tan or black interior, black grille, silver hubs, tires.
1. Yellow gold body.
2. Silver body.
3. Pale rose body.
4. Red body.
5. Brick red body.
6. Orange body.
7. Yellow body.
8. Light green body.
9. Green body.
10. Metallic green body.
11. Tan body.
12. Golden brown body.
13. Metallic gray body.

11   MERCEDES-BENZ 350 SL   (announced 1972, not issued)

12   PEUGEOT 104   82 mm   1/43   1/1973-1980>#1012
Hatchback sedan with cast body, matching opening front doors, plastic windows and lights, black chassis and grille, brown, gray or black interior,

silver bumpers and hubs, tires, suspension.
1. Red body.
2. Red-orange body.
3. Yellow body.
4. Yellow-green body.
5. Light green body.
6. Green body.
7. Light olive body.
8. Metallic silver blue body
9. Tan body.

12B PEUGEOT 104 DRIVING SCHOOL CAR 82 mm 1/43 2/1977-1980

Same basic model as #12, with decals or labels and roof sign.
1. Cream body.
2. Bright red body.

13 MATRA 670 SHORT 98 mm 1/43 3/1973-1977

Sports-racing car with cast body, black chassis, plastic lights, white cowling and air scoop, black seats and roll bar, wheels with spoked hubs, suspension, #15 and various decals.
1. Blue body.

14 MATRA 670 LONG 104 mm 3/1973-1978

Sports-racing car with cast body, black chassis, parts similar to #13, plus silver plastic airfoil, #14. Body has longer tails.
1. Blue body.

15 LOLA T280 82 mm 1/43 5/1973-1976

Sports-racing car with cast body, plastic lights, black chassis and seats, silver airfoil and parts. wheels with spoked hubs, suspension, #7 and various racing decals.
1. Yellow body.

16 FERRARI DAYTONA 98 mm 11/1973-1980>#1016

Sports coupe with cast body, matching opening doors, plastic windows and lights, black chassis-interior, wheels with spoked hubs. Some versions have racing decals.
1. White body, #118 and Thomson decals with red panels.
2. Red body, #54 and NART decals and stripes.
3. Yellow body.

17 GULF MIRAGE 85 mm 1/43 11/1973-1980>#1017

Sports-racing car with cast body, matching opening rear hood, chassis, plastic lights, orange air scoop, silver airfoil and gearbox, black seats, wheels with spoked hubs, suspension, #5, orange trim, Gulf and Ford decals.
1. Light blue body, orange chassis.
2. Light blue body, yellow-orange chassis.

18 PORSCHE 917/10 101 mm 1/43 7/1973-1980

Sports-racing car with cast front and rear body, chassis, white and/or greenish plastic parts, black seats, silver mirrors, wheels with spoked hubs, suspension, #7 and other racing decals with red and black stripes.
1. White body, silver chassis.

18B PORSCHE 917/10 101 mm 1/43 3/1974-1977

Same model as #18, with #2, Bosch and other decals.
1. Dark yellow body, red chassis.

18C PORSCHE 917/10 101 mm 1/43 1977-1980>#1018

Same model as #1018, with #11 and Uniroyal decals.
1. Red body and chassis.

19 ALFA ROMEO 33 TT (announced 1973, issued later as #41)

19 VOLKSWAGEN GOLF 85 mm 1/43 6/1975-1980>#1019

Two-door hatchback with cast body, matching opening doors, plastic windows, black chassis, interior and grille, silver bumpers and hubs, tires, suspension.
1. Red body.
2. Light yellow body.
3. Dark yellow body.
4. Dark yellow body, PTT decals.
5. Yellow-green body.
6. Light green body.
7. Olive green body.
8. Metallic blue body.

20 FERRARI 312 PB (announced 1973, not issued)

20 ALPINE RENAULT A441 99 mm 10/1975-1980>#1020

Sports-racing car with cast body, matching rear hood and chassis, plastic lights, black seats, wheels with spoked hubs, #4 and other racing decals with red and white stripes.
1. Blue body and chassis.

21 MATRA SIMCA BAGHEERA 90 mm 1/43 11/1973-1980>#1021

Fastback coupe with cast body, matching opening doors and lights, plastic windows, black chassis, interior and grille, silver hubs, tires, suspension.
1. Silver body.
2. Red body.
3. Orange-red body.
4. Light yellow body.
5. Dark yellow body.
6. Green body.

22 RENAULT 12 BREAK 101 mm 1/43 3/1975-1980>#1022

Wagon with cast body, matching opening hatch, plastic windows, black

(usually) chassis, interior and grille, silver bumpers and hubs, tires, suspension.
1. White body.
2. Orange body.
3. Metallic orange-gold body.
4. Yellow body.
5. Pale green body.
6. Metallic olive body.
7. Pale blue body, dark gray chassis.
8. Light blue body.
9. Blue body.
10. Bronze body.
11. Metallic gray body.

23 PEUGEOT 504 AMBULANCE 111 mm 1/43 4/1974-1980>#1123

Wagon with cast body, matching opening front doors and hatch, black chassis, plastic windows, blue dome light, tan interior, white stretcher, silver grille, bumpers and hubs, tires, suspension. red and white flag, blue "Ambulance Municipale" labels.
1. White body.

23B PEUGEOT 504 POLICE CAR 111 mm 1/43 12/1974-1980>#1124

Same basic model as #23, but with amber dome light, no stretcher, antenna in place of flag, may have "Gendarmerie" lettering.
1. Dark blue body.
2. Dark purplish-blue body.

23C PEUGEOT 504 BREAK 111 mm 1/43 2/1977-1980>#1125

Same basic model as #23, but no roof light, antenna, stretcher or lettering.
1. Silver body.
2. Green body.
3. Dark green body.

23D PEUGEOT 504 FIRE CAR 111 mm 1/43 2/1979-1980>#1126

Same basic model as #23, with antenna in place of flag, white "Sapeurs Pompiers" lettering.
1. Red body.

24 PORSCHE CARRERA RS 96 mm 1/43 5/1974-1980

Sports coupe with cast body and chassis, plastic windows and lights, black interior, silver exhaust pipes, fast wheels.
1. White body, red chassis and trim, #108 and other decals.
2. Light yellow body, otherwise as type 1.
3. Dark yellow body, otherwise as type 1.
4. Silver body and chassis, no decals.
5. Silver body, black chassis, no decals.
6. White body, red chassis, no decals.
7. Red body, black chassis, no decals.
8. Metallic tan body, black chassis, no decals.
9. Metallic gray body, black chassis, no decals.
10. Metallic gray body and chassis, no decals.

25 BMW 3.0 CLS 102 mm 1/43 7/1974-1980>#1025

Sports coupe with cast body, matching opening doors, black chassis, plastic windows, black interior and grille, wheels with silver hubs, #51, red-blue stripes and other decals.
1. White body.

26 FORD CAPRI 2600 RV 97 mm 1/43 10/1974-1980>#1026

Sports coupe with cast body, matching opening doors, plastic windows, black chassis-interior and grille, wheels with silver hubs, #55, white stripe and other decals.
1. Metallic light blue body.
2. Dark blue body.
3. Purplish-blue body.
4. Metallic purplish-blue body.

27 LANCIA STRATOS 85 mm 1/43 10/1974-1980>#1027

Sports-racing coupe with cast body, black chassis, plastic windows, black interior, wheels with spoked hubs, racing decals.
1. White body, red stripes and #4 decals.
2. White body, #111, white trim and Marlboro decals.
3. Red body, decals as type 1.
4. Red body, decals as type 2.
5. Darker red body, decals as type 2.

28 BMW 2002 TURBO 97 mm 1/43 11/1975-1980>#1028

Coupe with cast body, matching opening doors, black chassis, plastic windows, black interior and grille, silver bumpers, red and blue stripe decals.
1. Silver body.
2. White body.
3. Cream body.

29 LOLA GITANES (announced in 1974, not issued)

29 CITROEN CX 2200 107 mm 1/43 9/1975-1980>#1029

Sedan with cast body, matching opening front doors, black chassis, plastic windows and lights, tan (or other?) interior, black grille, silver bumpers, wheels with silver hubs, suspension.
1. Silver body.
2. White body.
3. Metallic red body.
4. Metallic pale blue body.
5. Metallic light bluye body.
6. Metallic blue body.
7. Metallic dark blue body.
8. Metallic golden-brown body.
9. Metallic red-brown body.

30 RENAULT 16 TX   (announced 1975, not issued)

30   RENAULT 30 TS   104 mm   1/43   12/1975-1980>#1030
Sedan with cast body, matching opening front doors, black chassis, plastic windows, tan interior, black grille, silver bumpers and hubs, tires, suspension.
1. Silver body.
2. Green body.
3. Metallic green body.
4. Metallic light blue body.
5. Metallic blue body.

31   DELAGE D8-120 1939   122 mm   1/43   12/1975-1980>#4031
Vintage convertible with cast body, chassis, plastic windshield, black or white raised top, black or tan interior, black grille, silver bumpers and other parts, wheels with wire hubs.
1. White body, metallic red chassis, black top, either interior.
2. White body, black chassis and top, tan interior.
3. Metallic red body, black chassis.
4. Wine red body, tan chassis.
5. Maroon body, tan chassis, white top, tan interior.
6. Green body, black chassis.
7. Light gray body, metallic red chassis.
8. Black body, metallic red chassis, white top, tan interior.

32   CITROEN 15 SIX 1939   110 mm   1/43   12/1974-1980>#4032
Sedan with cast body, matching opening hood sides, black chassis, plastic windows, gray interior, silver motor, lights, grille and bumpers, cream hubs, tires, cast-in rear spare.
1. Black body.

32A   CITROEN 15 SIX FIRE CAR   110 mm   1/43   2/1979-1980>#4033

Same model as #32, with red hubs, white "Sapeurs Pompiers" and badge decals.
1. Red body.

32B   CITROEN 15 SIX FFI   110 mm   1/43   6/1976-1980>#4034
Same model as #32, with thick or thin "FFI" and emblem decals.
1. Tan body with dark green camouflage.

33   FIAT X1/9   88 mm   1/43   7/1974-1980>#1033
Coupe with cast body, matching opening doors, plastic windows, black chassis, interior, engine cover and bumpers, silver hubs, tires, suspension.
1. Silver body.
2. Red body.
3. Orange body.
4. Metallic red body.
5. Green body.
6. Light blue body.

34   SIMCA 1100 TI   92 mm   1/43   7/1974-1976
Sedan with cast body, matching opening front doors and hatch, plastic windows, red or tan interior, black chassis, silver grille, bumpers and hubs, tires.
1. Silver body.
2. Red body.
3. Yellow body, red interior.
4. Yellow gold body.
5. Wine red body.
6. Metallic green body, tan interior.

35   DUESENBERG TYPE J 1931   130 mm   1/43   2/1976-1980>#4035
Vintage convertible with cast body, matching removable hood, chassis, plastic windshield, grayish top, light brown interior, silver motor, grille,

bumper and other parts, spoked whitewall tires, twin side spares.
1. White body, red chassis.
2. Red body, black chassis.
3. Red-orange body, white chassis.
4. Orange body, black chassis.
5. Orange body, black chassis.
6. Green body, black chassis.
7. Light blue body, black chassis.
8. Blue body, black chassis.
9. Light purplish-blue body, black chassis.
10. Dark purplish-blue body, ? chassis.
11. Tan body, white chassis.
12. Light gray body, black chassis.

36   PORSCHE 914/6 RALLY   91 mm   1/43   1/1975-1978
Coupe with cast body, matching opening light covers, plastic windshield, black top and chassis-interior, silver grille, rear bumper and hubs, tires, suspension, #40 and other decals.
1. White body.

37   RENAULT 17TS RALLY   96 mm   1/43   5/1975-1978
Hatchback with cast body, matching opening doors, plastic windows, black chassis-interior and grille, silver bumpers and hubs, tires, suspension, #6, Rallye du Maroc and other decals.
1. Blue body, red roof and cream trim decals.

38   GULF-FORD GR8   99 mm   1/43   2/1976-1980>#1038
Sports-racing car with cast body, chassis, plastic lights, black interior and exhaust pipes, silver mirrors, wheels with spoked hubs, #11, Gulf, orange trim and other decals.
1. Light blue body, orange chassis.

39   SIMCA 1308 GT   99 mm   1/43   3/1976-1980>#1039
Sedan with cast body, matching opening front doors, black chassis, plastic windows and lights, tan interior and bumpers, black dash and grille, silver hubs, tires, suspension.
1. Gold body.
2. Yellow gold body.
3. Wine red body.
4. Dark maroon body.
5. Metallic light green body.
6. Metallic green body.
7. Metallic light blue body.
8. Metallic red-brown body.
9. Metallic light golden brown body.
10. Metallic brown body.
11. Metallic silver gray body.
12. Metallic greenish-gray body.
13. Metallic bluish-gray body.
14. Metallic tannish-gray body.

40   PEUGEOT 604 V6 SL   107 mm   1/43   5/1976-1980>#1040
Sedan with cast body, matching opening hood and front doors, black chassis, plastic windows and lights, tan or black interior, black motor, silver motor area and hubs, tires, suspension.
1. Silver body.
2. Metallic maroon body.
3. Metallic dark green body, tan interior.
4. Metallic tan body.
5. Golden brown body.
6. Metallic brown body, tan interior.
7. Metallic gray body.
8. Metallic dark gray body.

41 ALFA ROMEO 33 TT 12 97 mm 1/43 6/1976-1980>#1041

Sports-racing car with cast body, matching rear hood with air scoop, black chassis, plastic windshield and lights, wheels with silver hubs, #1, Alfa Romeo, Campari and other decals.
   1. Red body.

42 RENAULT 4 VAN 88 mm 1/43 4/1976-1980>#1042

Minivan with cast body, matching opening rear door, plastic windows, cream chassis and hubs, black grille and interior, tires, suspension, logo decals or labels. Azema lists all types by letter except this one peomotional version:
   1. White body, blue "Projet" decals.

42A RENAULT 4 MAIL VAN 88 mm 1/43 6/1976-1980>#1001

Same model as #42, with dark blue or purple emblem decals.
   1. Yellow body.

42B RENAULT 4 FIRE VAN 88 mm 1/43 4/1976-1980>#1002

Same model as #42, with red hubs, Paris coat of arms decals.
   1. Red body.

42C RENAULT 4 SOLIDO VAN 88 mm 1/43 1976-1979

Same model as #42, with yellow-on-black Solido decals.
   1. Blue body.
   2. Dark blue body.

42D RENAULT 4 GAS VAN 88 mm 1/43 1977-1979

Same model as #42, with Gaz de France decals.
   1. Light blue body.

42E RENAULT 4 VAN 88 mm 1/43 1977-1980

Same model as #42, but no logo decals.
   1. White body.

42F RENAULT 4 CHAINEGAZ VAN 88 mm 1/43 1978-1980

Same model as #42, with Chainegaz decals.
   1. White body.
   2. Orange body.

42G RENAULT 4 CARAMBAR VAN 88 mm 1/43 8/1978

Same model as #42, with yellow and black Carambar decals.
   1. Reddish-purple body.

42H RENAULT 4 EDF/GDF VAN 88 mm 1/43 3/1979

Same model as #42, with EDF and GDF logo.
   1. Turquoise blue body.

43 RENAULT 14 92 mm 1/43 10/1976-1980>#1043

Sedan with cast body, matching opening front doors, plastic windows and lights, grayish-tan base, gray or black interior, black grille, silver wheels, tires.
   1. Silver body.
   2. Red body.
   3. Orange-red body.
   4. Yellow body.
   5. Metallic maroon body.
   6. Light green body.
   7. Green body.

44 FERRARI BB 100 mm 1/43 11/1976-1980>#1044

Sports coupe with cast body, matching opening rear hatch, black chassis, plastic windows (sometimes tinted), white interior, grille and lights, black airfoil, louvers and wipers, silver and black motor, spoked hubs, tires, suspension.
   1. Red body.
   2. Yellow body.

45 FORD ESCORT L 91 mm 1/43 7/1976-1980>#1045

Two-door with cast body, matching opening doors, plastic wheels, gray interior, black chassis and grille, silver hubs, tires, suspension.
1. Silver body.
2. Red body.
3. Yellow body.
4. Green body.

46 ROLLS-ROYCE PHANTOM III 1939 129 mm 1/43 12/1976-1980>#4046

Convertible with cast body, chassis, plastic windshield, black top, tan interior, silver grille, bumpers and hubs, tires, twin side spares.
1. Silver body, metallic blue chassis.
2. Silver body, gray chassis.
3. Silver body, black chassis.
4. Metallic pale green body, black chassis.
5. Metallic green body, black chassis.
6. Dark green body, black chassis.
7. Dark blue body, black chassis.

47 MERCEDES-BENZ 280E 107 mm 1/43 4/1977-1980>#1047

Sedan with cast body, matching opening hood and front doors, black chassis, plastic windows, yellow or tan interior, black parts, silver motor, grille, bumpers and hubs, tires, suspension.
1. Metallic silvery-gold body.
2. Off-white body.
3. Metallic red body.
4. Metallic dark red body.
5. Metallic green-gold body.
6. Metallic green body.
7. Metallic dark green body.
8. Metallic olive body.
9. Metallic dark olive body.
10. Metallic light blue body.
11. Metallic gray-blue body.
12. Metallic purple body.
13. Grayish-tan body.

48 DELAHAYE 135M 1937 117 mm 1/43 6/1977-1980>#1148>#4048

Convertible with cast body, chassis, plastic windshield, black or tan top, black, gray or tan interior, silver grille, bumpers and spoked hubs, tires, rear spare.
1. Red body and chassis, black top and interior.
2. Red body, black chassis.
3. Metallic red body and chassis.
4. Dark green body, black chassis, tan top and interior.
5. Light blue body, white chassis, black top and interior.
6. Light blue body, dark blue chassis.
7. Blue body, dark blue chassis, black top and interior.

49 PORSCHE 928 102 mm 1/43 7/1977-1980>#1049

Sports coupe with cast body, matching opening doors, black chassis, plastic windows and lights, red, tan or black interior, silver hubs, tires, suspension.
1. Silver body, black interior.
2. Light red body.
3. Red body, black or tan interior.
4. Metallic salmon body.
5. Metallic blue body, red interior.
6. Metallic dark blue body.
7. Dark blue body.
8. Metallic lilac-brown body.

50 PEUGEOT 504 RALLY

Sedan with cast body, black chassis, plastic windows and lights, tan

interior, black grille, silver bumpers and hubs, tires, suspension, #24 and rally decals.
1. White body.
2. White body, no decals.
3. Blue body (pre-series issue).

51 DELAGE D8 1939 COUPE DE VILLE 122 mm 1/43 9/1977-1980> #1151>#4051

Town car with cast body, chassis, plastic windows, tan interior, black grille, silver shell, bumpers, parts and spoked hubs, tires, twin side spares.
1. White body, dark gray chassis.
2. Green body, white chassis.
3. Rose-tan body, lilac-brown chassis.
4. tan body and chassis.
5. Tan body, brown chassis.

52 LANCIA BETA COUPE 92 mm 1/43 5/1977-1980>#1052

Coupe with cast body, matching opening doors, plastic windows, black chassis-interior, silver grille, bumpers and hubs, tires.
1. Orange body.
2. Wine red body.
3. Metallic pale green body.
4. Gray-green body.
5. Metallic olive body.
6. Brown body.

53 FORD FIESTA 83 mm 1/43 11/1977-1980>#1053

Two-door hatchback with cast body, matching opening doors, plastic windows, black chassis, interior and grille, silver bumpers and hubs, tires.
1. Silver body.
2. Red body.
3. Dark blue body.

54 FIAT 131 ABARTH RALLY 97 mm 1/43 12/1977-1980>#1054

Two-door with cast body, chassis, plastic windows, black interior, grille and airfoil, silver lights, wheels with spoked hubs, +/- #5, yellow roof panel and rally decals.
1. White body, red chassis, no decals.
2. Dark blue body, yellow chassis, decals.

55 CORD L29 1930 116 mm 1/43 11/1977-1980>#4055

Convertible sedan with cast body, matching hood, black chassis, plastic windows, black or gray top, black interior, silver grille, bumpers and wire hubs, usually whitewall tires.
1. Wine red body, black or gray top.
2. Yellow body, black top.
3. Green body, black top.
4. Dark blue body, black top.
6. Tan body, black top.

56 CITROEN 2CV6 88 mm 1/43 12/1977-1980>#1056

Sedan with cast body, matching lights, plastic windows, gray chassis, roof, interior and hubs, tires, suspension.
1. Orange body.
2. Light green body.

57 ALPINE A442 TURBO 102 mm 1/43 3/1977-1980>#1057

Sports-racing car with cast body, matching rear hood and air scoop, black chassis-interior, plastic lights, silver motor, wheels with spoked hubs, #5 and Alpine-Renault decals.
1. Yellow body.

58 RENAULT 5 GORDINI 78 mm 1/43 3/1977-1980>#1058

Hatchback with cast body, matching opening doors, plastic windows and lights, black or tan chassis, red interior, black grille and dash, silver hubs,

tires, suspension, #1 and rally decals.
1. Yellow body, black chassis.
2. Blue body, tan chassis.

58B RENAULT 5 ALPINE (announced 1977, not issued)

59 RENAULT 40CV 1926 123 mm 1/43 3/1977-1980>#1058>#4058

Phaeton with cast body, matching hood, black chassis, plastic windshield, light gray top, dark gray interior and trunk, silver lights and spoked hubs, tires, twin side spares, +/- hood emblem decal.
1. Red body.
2. Yellow body.
3. Green body.
4. Tan body.

60 SIMCA 1308 GT TAXI 97 mm 1/43 4/1977-1980>#1060

Sedan with cast body, matching opening front doors, black chassis, plastic windows, tan interior and bumpers, black grille and dash, gray roof sign, silver hubs, tires, suspension, black-on-yellow Taxi Radio labels on sides and roof sign.
1. White body.
2. Off-white body.
3. Light green body.
4. Green body.
5. Pale gray body.

61 FORD ESCORT RALLY 92 mm 1/43 6/1977-1980>#1061

Two-door with cast body, matching opening doors, plastic windows, black chassis, interior and grille, silver bumpers, lights and hubs, tires, suspension, #15, black panels, red stripes, Allied Polymer and other decals.
1. White body.

62 HISPANO-SUIZA H6B 1926 114 mm 1/43 10/1977-1980>#1162> #4062

Phaeton with cast body, chassis, unpainted drive train, plastic windshield, black top, grille, trunk and side spares, brown interior, silver lights, shell and hubs, tires.
1. Metallic green body, black chassis.
2. Light blue-gray body, blue chassis.
3. Blue-gray body, dark blue chassis.
4. Gray-brown body, white chassis.

63 PORSCHE 911 TURBO 98 mm 1/43 1/1987-1980>#1063

Sports coupe with cast body, matching opening doors, black chassis with "Porsche 930" lettering, plastic windows and lights, black interior and airfoil, wheels with spoked hubs.
1. Silver body.
2. Orange body.
3. Yellow-orange body.
4. Metallic light green body.
5. Metallic green-gray body.
6. Metallic blue-gray body.

64 RENAULT 4 EDF VAN (announced 1974, not issued)

65 CITROEN CX 2400 BREAK 102 mm 1/43 6/1978-1980>#1065

Wagon with cast body, matching opening hatch, black chassis, plastic windows and lights, tan or brown interior, black grille, silver bumpers, wheels with sliver hubs.
1. Metallic blue body.
2. Metallic blue-gray body.
3. Metallic dark blue body.
4. Metallic purple body.
5. Pale grayish-tan body.

66 LAND ROVER 109   106 mm   1/43   6/1978-1980>#1066

Closed car with cast body and chassis, plastic windows, roof and opening rear door matching body, black interior, gray hubs, tires, rear spare, suspension.
1. Dull blue body and chassis.
2. Tan body and chassis.
3. Olive tan body and chassis.

67 MERCEDES-BENZ 540K 1939   122 mm   1/43   5/1978-1980>#4067

Convertible with cast body, chassis, plastic windshield, top, interior, silver grille, bumpers and wire hubs, tires.
1. Silver body and chassis, black top and interior.
2. Metallic red body and chassis, white top, tan interior.
3. Maroon body and chassis, white or black top, tan interior.
4. Light brown body, brown chassis, light tan top, tan interior.

68 PORSCHE 934 TURBO   99 mm   1/43   1/1979-1980>#1968

Same model as #63, with decals, "Porsche 934 Turbo" on base.
1. White body, #69, white trim, VSD and other decals.
2. White body, #86, red and black trim and other decals.
3. Red body, same decals as type 1.

69 ALFASUD TROPHEE   87 mm   1/43   9/1978-1980>#1069

Sports coupe with cast body, plastic windows, black chassis and grille, gray or black interior, silver hubs, tires, #25, Alfasud and other decals.
1. Red body.
2. Purple body.

70 OPEL KADETT GTE RALLY   96 mm   1/43   10/1978-1980>#1070

Coupe with cast body, plastic windows, black chassis-interior, silver lights and hubs, tires, #31, Panta Shop and other decals.
1. Orange body.
2. Orange-red body.

71 ROLLS-ROYCE PHANTOM III 1939   130 mm   1/43   9/1978-1980>#4071

Town car with cast body, chassis, plastic windows, tan, olive or dark gray interior, silver grille, bumpers and hubs, tires, twin side spares.
1. Pale yellow body, black chassis.
2. Brownish-green body, tan chassis.
3. Brownish-green body, grayish-tan chassis.
4. Brownish-green body, olive green chassis.
5. Olive green body, brownish-green chassis.
6. Dark blue body, light blue chassis.
7. Light gray body, dark gray chassis.
8. Dark gray body, light gray chassis.
9. Tannish-gray body, brown chassis.
10. Brown body, black chassis.
11. Tan body, black chassis.
12. Pale tan body, brown chassis.
13. Rosy-tan body, brown chassis.

72 CITROEN LN   76 mm   1/43   3/1978-1980>#1072

Hatchback with cast body, matching opening hatch, plastic windows, black chassis-interior and grille, silver bumpers and hubs, tires.
1. White body.
2. Light blue body.
3. Blue body.
4. Dark blue body.

73 LANCIA STRATOS   85 mm   1/43   1/1979-1980>#1073

Sports coupe with cast body, black chassis, plastic windows, black interior and louvers, wheels with yellow spoked hubs, #10 and Monte Carlo Rally decals.
1. Blue body.

73B LANCIA STRATOS 85 mm 1/43 1/1979-1980>#1074

Same model as #73, with #19 and Tour de Corse decals.
1. White body.

75 BMW 3000 CSL 105 mm 1/43 7/1978-1980>#1075

Sports coupe with cast body, matching opening doors, black chassis, plastic windows, black interior and grille, wheels with silver hubs, #21, Jägermeister and other decals.
1. Light orange body.
2. Dark orange body.

76 SIMCA CHRYSLER HORIZON 92 mm 1/43 11/1978-1980>1076

Hatchback sedan with cast body, matching opening front doors, black chassis, plastic windows, tan or black interior, black grille, silver bumpers and hubs, tires, suspension.
1. Red body, black interior.
2. Metallic light green body.
3. Metallic green body.
4. Dark brown body, tan interior.
5. Metallic gray body.

77 ROLLS-ROYCE PHANTOM III 1939 130 mm 1/43 2/1978-1980>#4077

Open convertible with cast body, chassis, plastic windshield, tan or brown interior, black folded top, silver grille, bumpers and hubs, tires, twin side spares.
1. Off-white body and chassis.
2. Cream body and chassis.
3. Cream body, black chassis.
4. Gray-brown body, black chassis.

78 DELAHAYE 135M 1939 118 mm 1/43 4/1978-1980>#4078

Open convertible with cast body, chassis, plastic windshield, black interior and folded top, silver grille, bumpers and wire hubs, tires, rear spare.
1. Red body and chassis.
2. Red body, black chassis.
3. Metallic red body and chassis.
4. Dark green body, black chassis.
5. Light blue body, white chassis.
6. Light blue body, dark blue chassis.

80 CORD L29 1929 COUPE 117 mm 1/43 2/1979-1980>#4080

Coupe with cast body, chassis, plastic windshield, tan interior except type 5, white, gray or black top, silver grille, bumpers and wire hubs, whitewall tires.
1. Red body, black chassis, white top.
2. Wine red body, black chassis, gray or black top.
3. Dark blue body and chassis, black top.
4. Dark blue body, black chassis, gray or black top.
5. Pale tan body, brown chassis and interior, white top.

81 PEUGEOT 104 ZS 77 mm 1/43 12/1978-1980>#1081

Same basic model as #72, with #59, Rallye d'Antibes and other decals.
1. Mustard tan body and hatch.
2. Greenish yellow body and hatch.
3. Black body and hatch.

82 ALFA ROMEO ALFETTA GTV 96 mm 1/43 3/1979-1980>#1082

Fastback coupe with cast body, black chassis, plastic windows, black interior and grille, wheels with silver hubs, #31, black hood, Garden Stores and other decals.
1. White body.

85 CADILLAC 452A 1930 129 mm 1/43 10/1979-1980>#4085

Landaulet with cast body, chassis, plastic windows, black or white roof and open rear top, gray-brown interior, silver grille, bumpers and wire hubs, whitewall tires.
1. Off-white body, black chassis, white roof.
2. Light green body, green chassis, white roof.
3. Olive green body, green chassis, white roof.
4. Blue body, dark blue chassis, white roof.
5. Light tan body, black chassis and roof.
6. Gray-brown body, black chassis and roof.

86 PORSCHE 936 LE MANS 108 mm 1/43 4/1979-1980>#1086

Sports-racing car with cast body, plastic lights, black chassis-interior, white air scoop and hubs, wheels, #4, red and blue stripes, Martini Porsche and other decals.
1. White body.

87 ALPINE RENAULT A442B 110 mm 1/43 5/1979-1980>#1087

Sports-racing car with cast body, black chassis, plastic windshield and lights, yellow air scoop, wheels with gray hubs, #2, black and white trim, Renault elf and other decals.
1. Light yellow body.
2. Yellow body.

88 BUGATTI ATALANTE 1939 COUPE 109 mm 1/43 5/1979-1980>#4088

Sports coupe with cast body, chassis, plastic windows, light gray (or tan?) interior, silver grille, bumpers and wire hubs, tires, rear spare, grille emblem label.
1. White body and chassis.
2. Maroon body and chassis.
3. Light blue body and chassis.
4. Light blue body, blue chassis.
5. Blue body and chassis.
6. Black body and chassis.

89 BMW 530 103 mm 1/43 7/1979-1980>#1089

Sedan with cast body, black chassis, plastic windows, tan or black interior, black dash and grille, silver hubs, tires, BMW emblem, red and blue stripes and other decals.
1. Silver body.
2. White body.
3. Metallic light blue body.
4. Metallic dark blue body.

90 PEUGEOT 305 97 MM 1/43 6/1979-1980>#1090

Sedan with cast body, black chassis, plastic windows, tan or dark brown interior, black dash, grille and bumpers, silver hubs, tires, suspension.
1. Cream body.
2. Metallic orange-red body.
3. Metallic light orange-red body.
4. Metallic red body.

91 RENAULT 18 100 mm 1/43 12/1978-1980>#1091

Sedan with cast body, matching opening front doors, black chassis, plastic windows, tan interior, black dash and grille, silver lights, bumpers and hubs, tires, suspension.
1. Metallic yellow-green body.
2. Metallic light green body.
3. Metallic green body.
4. Blue body.
3. Dark blue body.

94 TOYOTA CELICA (announced 1979, issued 1980 as #1094)

96 JAGUAR XJ 12   110 mm   1/43   1/1980>#1096
Sedan with cast body, matching opening front doors, black chassis, plastic windows and lights, tan interior, silver grille, bumpers and hubs, tires, suspension.
   1. Silver body.
   2. White body.
   3. Red body.
   4. Maroon body.
   5. Dark green body.
   6. Dark blue body.
   7. Blue-black body.

97 RENAULT REINASTELLA 1934   122 mm   1/43   1/1980>#4097
Sedan with cast body, chassis, plastic windows, gray interior, silver grille and bumpers, black wheels, silver hubcaps, twin side spares.
   1. Dark blue body and chassis.
   2. Light gray body, gray chassis.
   3. Pale gray body, blue-gray chassis.
   4. Gray body, dark gray chassis.
   5. Black body and chassis.

Note: Several 10 and 100 series models were offered in kit form in the late seventies. These were numbered with the model's usual catalog number suffixed with the letter K. In 1980 they were made into the 5000 series, where they are listed in this book. No changes in their colors or decals were made when the numbers were changed, and the 5000 series listings cite their earlier numbers.

AIRCRAFT

   Solido introduced a series of diecast aircraft in 1955, and added models to the line in the next two years. The series remained in production until at least 1960.

163 MYSTERE IV A   83 x 74 mm   1/150   1955
Jet fighter, single casting with blue canopy, French roundels, plus 3 wheels.
   1. Silver.

164 FOUGA MAGISTER CM-170   63 x 74 mm   1/150   1955
Jet plane, single casting, with or without wing tanks, with French roundels, plus 3 wheels.
   1. Silver.

165 SKYRAY   84 x 66 mm   1/150   1955
Delta-wing jet, single casting with blue or silver canopy, US Navy or Air Force markings, plus 3 wheels.
   1. Silver (USAF).
   2. Blue (USN).
   3. Metallic blue (USN).
   4. Dark blue (USN).

166 VAUTOUR   1/150   1955
Jet, single casting, with blue canopy and French roundels, plus three wheels.
   1. Silver.

167 LEDUC 021   82 x 78 mm   1/150   1955
Jet, single casting with blue nose, French roundels.
   1. Silver.
   2. Pale green.

168 SUPER-SABRE F100   90 x 72 mm   1/150   1955
Jet fighter, single casting with blue canopy, USAF markings, plus three wheels.
   1. Silver.

169 SIKORSKY S-55 HELICOPTER   1/150   1955

Helicopter, body casting with various markings, float casting, and two rotors.
1. Silver, with red crosses.
2. White, with red crosses.
3. Dark blue (navy).
4. Blue-gray (civilian).
5. Olive, with French roundels (army).

170 MIG-15 75 x 72 mm 1/150 1956
Jet fighter, single casting with blue canopy and red stars, plus three wheels.
1. White.

171 HAWKER HUNTER 1/150 1956
Jet fighter, single casting, with gray canopy and British roundels, plus three wheels.
1. Gray and green camouflage.

172 THUNDERJET 68 x 74 mm 1/150 1956
Jet fighter, single casting with blue canopy and US markings, plus three wheels.
1. Silver.

173 BARODEUR 95 x 67 mm 1/150 1956
Jet fighter, body casting with blue canopy and French roundels, plus gray landing sled with three wheels.
1. Silver.

174 JAVELIN 1/150 1956
Delta-wing jet, single casting with blue or gray canopy and British roundels, plus three wheels.
1. Pale green and green camouflage.
2. Light gray and green camouflage.
3. Dark gray and green camouflage.

175 SUPER-CONSTELLATION 1/300 1956
Four-propeller airliner, single casting? Two versions with TWA logo; there may be other types. Four propellers and ? wheels.
1. Silver, with red TWA logo and stripes.
2. Metallic blue and white, with red TWA logo.

176 CARAVELLE 1/300 1956
Jet airliner, fuselage and wing castings, plus wheels.
1. Silver and white, Air France logo.
2. Light blue and white, Air France logo.
3. Finnair logo (#176-1).
4. Alitalia logo (#176-2).
5. Silver and white, Air Maroc logo (#176-3)
6. Sud-Aviation logo (#176-6).
7. Silver and white, Varig logo (#176-8).
8. Silver and white, Swissair logo (#176-13).
9. Silver and white, United logo (#176-14).
10. Air Algérie logo (#176-15).
11. S.A.S. logo (#176-16).
12. Unknown other versions.

177 AUBERT SUPER-CIGALE 47 x 64 mm 1/150 1956
Light plane, single casting with silver canopy, plus propeller.
1. Silver.
2. White, with red cross.
3. Cream.
4. Red.
5. Yellow.
6. Light blue.

178 TRIDENT 87 x 51 mm 1/150 1956
Jet fighter, single casting with French roundels, plus three wheels.
1. Silver.

179 FAIREY DELTA 90 x 53 mm 1/150 1957
Delta-wing jet, single casting with blue canopy and British roundels, plus three wheels.
  1. Silver.

180 CONVAIR XFY-1 59 x 56 mm 1/150 1957
VTOL plane, body and nose castings with blue canopy and US markings, plus two propellers.
  1. Silver.

181A PIASECKI H-21 HELICOPTER 104 mm 1/150 1957
Helicopter, single casting with blue windows and French roundels, plus two rotors and three wheels. Piaseki (sic) name cast in.
  1. Silver.
  2. Green.

181B VERTOL H-21 HELICOPTER 1/150 1957
Same model as 181A, with Vertol name cast in. Same details.
  1. Silver.
  2. Green.

182 MORANE SAULNIER PARIS 65 x 67 mm 1/150 1957
Jet, single casting with blue canopy and French roundels, plus three wheels.
  1. Silver.

183 AQUILON 76 x 87 1/150 1957
Twin-tailed jet, single casting with silver canopy and French roundels, plus three wheels.
  1. Metallic blue.

184 SUPER-MYSTERE B2 92 x 69 mm 1/150 1957
Jet fighter, single casting with blue canopy and French roundels, plus three wheels.
  1. Silver.

185 ÉTENDARD IV 89 x 64 mm 1/150 1957
Jet fighter, single casting with silver canopy and French roundels, plus three wheels.
  1. Metallic blue.

186 BRÉGUET ALIZÉ 1/150 1957
Prop-jet?, single casting with silver canopy, French roundels and flag, plus propeller and three wheels.
  1. Metallic blue.

187 TUPOLEV 104 125 x 123 mm 1/300 1957
Jet airliner, fuselage and wing castings, with red stripe, blue windows and fuselage Aeroflot and CCCP number lettering, black wing lettering, and ten wheels.
  1. White.

POST-1980 MODELS

THE 1000 SERIES

When the new four-digit numbering system was instituted in 1980, the surviving 10 series models became the basis of the 1000 series by having 10 prefixed to their two-digit catalog numbers. A few models were given numbers not related to their previous numbers, and a few new models were added to the 1000 series before it was replaced by the 1300 series in 1982.

1001 RENAULT 4 MAIL VAN 88 mm 1/43 #42A>1980-1981

1002 RENAULT 4 FIRE VAN 88 mm 1/43 #42B>1980-1981

1004 RENAULT 4 CHAINEGAZ VAN 88 mm 1/43 #42F>1980

1005 CITROEN VISA (issued as #1302)

1006 FIAT RITMO (issued as #1303)

1007 CITROEN 2CV PETITE FLEUR (issued as 1301)

1010 RENAULT 5 78 mm 1/43 #10>1980-1982

1012 PEUGEOT 104 82 mm 1/43 #12>1980-1982

1016 FERRARI DAYTONA 98 mm 1/43 #16B>1980-1981

1017 GULF MIRAGE 85 mm 1/43 #17>1980

1018 PORSCHE 917/10 101 mm 1/43 #18c>1980-1981

1019 VOLKSWAGEN GOLF 85 mm 1/43 #19>1980-1981

1020 ALPINE RENAULT A441 99 mm 1/43 #20>1980

1021 MATRA SIMCA BAGHEERA 90 mm 1/43 #21>1980

1022 RENAULT 12 BREAK 101 mm 1/43 #22>1980-1981

1023 RENAULT 5 TURBO 83 mm 1/43 9/1981-1982>#1321

Hatchback with cast body, plastic windows and lights, light blue chassis-seats, orange rear interior, black grille and dash, wheels with spoked hubs, #18 and Tour de Corse decals.
  1. Red body.
  2. Metallic light blue body.
  3. Metallic blue body.
  4. Dark blue body.

1025   BMW 3.0 CSL   102 mm   1/43   #25>1980-1981

1026   FORD CAPRI 2600 RV   97 mm   1/43   #26>1980-1981

1027   LANCIA STRATOS   85 mm   1/43   #27>1980

1028   BMW 2002 TURBO   97 mm   1/43   #28>1980-1981

1029   CITROEN CX 2200   107 mm   1/43   #29>1980-1981

1030   RENAULT 30 TS   104 mm   1/43   #30>1980-1981

1031   BMW M1 PROCAR   102 mm   1/43   4/1981-1982>#1329
Sports-racing coupe with cast body, plastic windows, black chassis-interior, louvers and grille, wheels with conical hubs, #41 and Air Press decals.
  1. Light blue body.
  2. Dark blue body.
  3. Purplish-blue body.

1032   PORSCHE 935 LE MANS   108 mm   1/43   7/1980-1982
Sports-racing coupe with cast body, plastic windows, black chassis-interior-grilles, wheels with conical hubs, +/- decals.
  1. Red body, yellow hubs, #70, Paul Newman and other decals.
  2. Black body, no decals?

1033   FIAT X1/9   88 mm   1/43   #33>1980-1981

1034   FIRE LAND-ROVER   106 mm   1/43   5/1980-1982>#2104
Closed vehicle with cast body and chassis, plastic windows and lights, white roof, red opening door, tow hook and hubs, tires, spare wheel, white Sapeurs Pompiers and badge decals.
  1. Red body and chassis.

1038   GULF FORD GR8   99 mm   1/43   #38>1980

1039   SIMCA 1308 GT   99 mm   1/43   #39>1980-1981

1040   PEUGEOT 604   107 mm   1/43   #40>1980-1982

1041   ALFA ROMEO 33TT   97 mm   1/43   #41>1980

1042   RENAULT 4 SOLIDO VAN   88 mm   1/43   #42C>1980-1982>#1325

1043   RENAULT 14   92 mm   1/43   #43>1980-1981

1044   FERRARI BB   100 mm   1/43   #44>1980-1981

1045   FORD ESCORT   91 mm   1/43   #45>1980-1981

1047   MERCEDES-BENZ 280   107 mm   1/43   #47>1980-1982

1049   PORSCHE 928   102 mm   1/43   #49>1980-1981

1050   PEUGEOT 504 RALLY   103 mm   1/43   #50>1980-1982

1051   PORSCHE 924 TURBO   96 mm   1/43   3/1980-1982

Sports coupe with cast body, matching opening doors, black chassis, plastic windows and lights, red or tan interior, black dash and grille, spoked hubs, tires, suspension.
1. Dark green body.
2. Charcoal gray body.

1052 LANCIA BETA COUPE 92 mm 1/43 #52>1980-1981

1053 FORD FIESTA 83 mm 1/43 #53>1980-1981

1054 FIAT 131 ABARTH RALLY 97 mm 1/43 #54>1980-1982

1055 PEUGEOT 504 V6 COUPE 99 mm 1/43 3/1980-1982
Coupe with cast body, black chassis, plastic windows and lights, black interior and grille, silver bumpers and hubs, tires, suspension, $02, red-yellow-blue trim and other rally decals.
1. White body.

1056 CITROEN 2CV6 88 mm 1/43 #56>1980-1981

1057 ALPINE A442 TURBO 102 mm 1/43 #57>1980-1981

1058 RENAULT 5 GORDINI 78 mm 1/43 #58>1980-1981

1059 VOLKSWAGEN SCIROCCO 87 mm 1/43 4/1980-1982
Hatchback with cast body, black chassis, plastic windows, red interior, silver hubs, tires, suspension, #266 and other decals.
1. Black body.

1060 SIMCA 1308 GT TAXI 97 mm 1/43 #60>1980-1981

1061 FORD ESCORT RALLY 92 mm 1/43 #61>1980-1981

1062 MATRA RANCHO 98 mm 1/43 1/1981-1983>#2004
Wagon with cast body, plastic windows and opening hatch, black chassis, interior, tailgate and parts, silver hubs, tires.
1. Red body.
2. Red-gold body.
3. Yellow body.
4. Golden brown body.

1063 PORSCHE 911 TURBO 98 mm 1/43 #63>1980

1065 CITROEN CX 2400 BREAK 112 mm 1/43 #65>1980-1981

1066 LAND-ROVER 109 106 mm 1/43 #66>1980-1981

1068 PORSCHE 934 TURBO 99 mm 1/43 #68>1980-1982

1069 ALFASUD TROPHEE 87 mm 1/43 #69>1980-1981

1070 OPEL KADETT GTE RALLY 96 mm 1/43 #70>1980-1982

1072 CITROEN LN 76 mm 1/43 #72>1980-1981

1073 LANCIA STRATOS MONTE CARLO 85 mm 1/43 #73>1980-1982

1074 LANCIA STRATOS TOUR DE CORSE 85 mm 1/43 #73B>1980-1982

1075 BMW 3000 CSL 105 mm 1/43 #75>1980-1982

1076 SIMCA CHRYSLER HORIZON 92 mm 1/43 #76>1980-1981

1081 PEUGEOT 104 ZS 77 mm 1/43 #81>1980-1981

1082 ALFA ROMEO ALFETTA GTV 96 mm 1/43 #82>1980-1981

1086 PORSCHE 936 LE MANS 108 mm 1/43 #86>1980-1982

1087 ALPINE RENAULT A442B 110 mm 1/43 #87>1980-1982

1089 BMW 530 103 mm 1/43 #89>1980-1982

1090 PEUGEOT 305 97 mm 1/43 #90>1980-1981

1091 RENAULT 18 100 mm 1/43 #91>1980-1981

1094 TOYOTA CELICA 96 mm 1/43 4/1980-1982
Coupe with cast body, black chassis, plastic windows, black interior and grille, silver lights and hubs, tires, suspension, #29, Acropolis Rally and other decals.
 1. Cream body.

1096 JAGUAR XJ 12 110 mm 1/43 #96>1980-1982

1097 PORSCHE 934 LE MANS 99 mm 1/43 #68>1980
Same model as #68, type 2, also found minus decals.
 1. White body.

THE 1100 SERIES

The few surviving members of the 100 series, the four versions of the #23 Peugeot 504 wagon, and four Age d'Or oldtimers from the 10 series were put into the 1100 series in 1980. The modern cars were all deleted by the end of 1981, The four Golden Agers moved to the 4000 series during 1980--in fact, Azema does not list them, and I would not have known of their existence if I had not bought them in boxes numbered with 1100 numbers. True, a label with the name and number was glued onto each box, which may have been done by an importer rather than by Solido itself--but somebody went to the trouble of printing gummed labels, and these four models (and there could have been more) were marketed under these numbers for a short time.

1123 PEUGEOT 504 AMBULANCE 111 mm 1/43 #23A>1980-1981
Same model as #23A.
 1. White body.

1124 PEUGEOT 504 POLICE 111 mm 1/43 #23B>1980
Same model as #23B.
 1. Dark blue body.

1125 PEUGEOT 504 BREAK 111 mm 1/43 #23C>1980-1981
Same model as #23C. Standard colors.

1126 PEUGEOT 504 FIRE 111 mm 1/43 #23D>1980-1981
Same model as #23D.
 1. Red body.

1148 DELAHAYE FIGONI 1939 117 mm 1/43 #48>1980>#4048
Same model as #48.
 1. Red body, black chassis, top and interior.

1151 DELAGE 1939 COUPE DE VILLE 122 1/43 #51>1980>#4051
Same model as #51.
  1. Tan body and chassis, light tan interior.

1159 RENAULT 40CV PHAETON 123 mm 1/43 #59>1980>#4059
Same model as #59.
  1. Yellow body, black chassis, gray-brown roof.

1162 HISPANO-SUIZA 1926 114 mm 1/43 #62>1980>$4062
Same model as #62.
  1. Blue-gray body, dark blue chassis, black roof, brown interior.

1181 ALPINE 1600 90 mm 1/43 #181>1980-1981
Same model as #181. Standard colors.

1192 ALPINE RENAULT A310 96 mm 1/43 #192>1980
Same model as #192. Standard colors.

1193 CITROEN GS 95 mm 1/43 #193>1980-1981
Same model as #193. Standard colors.

1194 ALPINE RENAULT A310 POLICE 96 mm 1/43 #192B>1980
Same model as #192B.
  1. Blue body.

1196 RENAULT 17 TS 96 mm 1/43 #196>1980
Same model as #196. Standard colors.

1197 FERRARI 512M SUNOCO 98 mm 1/43 #197>1980
Same model as #197.
  1. Metallic purplish-blue body.

1198 PORSCHE 917K 96 mm 1/43 #198>1980-1981
Same model as #198. Standard colors.

THE 1200 SERIES

These models were made in Portugal in 1988 and 1989, and marketed at low prices as "Solido Toys". They are essentially reissues of 1300 series models, and almost all were issued in two different versions, known as A and B, which will be listed in that order below (1 = A, 2 = B).

1201 CITROEN VISA 85 mm 1/43 #1302>1988-1989
Same model as #1302.
  1. Maroon body.
  2. White body, +/- Citroen and Total logo.

1202 FIAT RITMO 89 mm 1/43 #1303>1988-1989
Same model as #1303. Black chassis and interior, fast wheels.
  1. White body, red and green trim, Avia logo.
  2. White body, black stripes, #64.

1203 RENAULT FUEGO 99 mm 1/43 #1308>1988-1989
Same model as #1308.
  1. Blue body, Gendarmerie logo.
  2. Yellow body, Dinin logo.

1204 PORSCHE 934 100 mm 1/43 #1323>1988-1989
Same model as #1323.
  1. Metallic brown body.
  2. White body, red and black stripes, Bridgestone logo, #66.
  3. Light yellow body, Royal Canin logo (promo).

1205 LANCIA RALLY 89 mm 1/43 #1327>1988-1989
Same model as #1327.
  1. Red body, Motul logo.
  2. Blue body, +/- Chardonnet logo.

1206 FORD SIERRA XR4 101 mm 1/43 #1340>1988-1989
Same model as #1340.
1. Maroon body.
2. Black body, +/- rear wing, Texaco logo. Promo.

1207 FORD ESCORT RS TURBO 91 mm 1/43 #1315>1988-1989
Same model as #1315, with black chassis-interior, grille and bumpers.
1. Red body.
2. Yellow body, black trim, Motul logo, #59.
3. Blue body.

1208 RENAULT 5 MAXI TURBO 82 mm 1/43 #1321>1988-1989
Same model as #1321.
1. Metallic gray body, Motul logo.
2. Red body, Opel Renault Sport logo.

1209 BMW M1 104 mm 1/43 #1329>1988-1989
Same model as #1329.
1. Red body.
2. White body, red, light and dark blue stripes, #11.
3. Light yellow body, Royal Canin logo. Promo.

1210 CITROEN 2CV 88 mm 1/43 #1301>1988
Same model as #1301.
1. Yellow body, Echappement logo.
2. White body, orange and yellow stripes, #2.
3. White red, blue, green and gray set of promos.

1211 RENAULT SUPER 5 86 mm 1/43 #1353>1988-1989
Same model as #1315.
1. Metallic blue body.
2. Yellow body.

3. Red body, Pompiers logo (fire car).
4. White body, Roiret logo.

1212 VOLKSWAGEN GOLF GTI 85 mm 1/43 #1314>1988-1989
Same model as #1314.
1. Black body, +/- Monroe logo.
2. White body, +/- Olivetti logo.
3. White body, Ivry la Bataille logo. Promo.
4. Metallic dark gray body.

1213 RENAULT 4 VAN 89 mm 1/43 #1325>1989
Same model as #1325, with flashing light. x
1. Red body.

1214 PORSCHE 935 108 mm 1.43 #1332>1989
Same model as #1332. No other data.

1215 AUDI QUATTRO 102 mm 1/43 #1328>1989
Same model as #1328. No other data.

1216 NISSAN PRAIRIE 93 mm 1/43 #1341>1989
Same model as #1341. No other data.

1217 BMW 530 104 mm 1/43 #1304>1989
Same model as #1304. No other data.

THE 1300 COUGAR SERIES

This was a special series of 1300 models produced from 1980 to 1982 and sold in bubblepacks. The first six of them, numbered 1401 to 1406, were also produced for sale as Dinky Toys and bear the 1400 number and Dinky Toys name on their chassis. In 1982 they became the first models in the Solido 1300 series, and their chassis lettering was changed accordingly.

1301 CITROEN 2CV6 88 mm 1/43 1980-1982
Sedan with cast body, matching lights, plastic windows, open or closed roof, interior, gray chassis-grille, fast wheels, yellow duck decals.
1. Orange body, light orange interior, tan open roof.
2. Green body, gray interior and closed roof.

1302 CITROEN VISA 85 mm 1/43 1980-1982
Sedan with cast body, plastic windows and lights, interior, chassis and matching grille, fast wheels.
1. Metallic red body, gray interior, chassis.
2. Metallic green body, yellow-green interior, white chassis.

1303 FIAT RITMO 89 mm 1/43 1980-1982
Hatchback with cast body, plastic windows and lights, chassis and interior, fast wheels.
1. Metallic blue body, yellow chassis and interior.
2. Metallic bronze body, light tan chassis and interior.

1304 BMW 530 104 mm 1/43 1980-1982
Sedan with cast body, tinted plastic windows, gray chassis and interior, black dash and grille, fast wheels, decals.
1. Metallic green body, black cougar decals.
2. Metallic light green body, black cougar decals.
3. Metallic purple body, red flame decals.

1305 ALFA ROMEO ALFETTA GTV 93 mm 1/43 1980-1982
Fastback with cast body, tinted plastic windows, tan chassis and interior, black dash and grille, fast wheels, clover decals.
1. Red body.
2. Yellow body.

1306 PEUGEOT 504 103 mm 1/43 1980-1982
Sedan with cast body, plastic lights, chassis, interior, tinted windows, black grille, silver bumpers, fast wheels, decals.
1. Metallic yellow gold body, brown chassis and interior, black cougar decals.
2. Metallic blue body, light blue chassis and interior, red flame   decals.

1307 TALBOT TAGORA 106 mm 1/43 1981-1982
Details as #1307 below.
1. Metallic tan body.
2. Metallic gray body.

1308 RENAULT FUEGO (not issued)

1309 RENAULT 14 92 mm 1/43 1981-1982
Details as #1309 below.
1. Metallic green body.
2. Metallic blue body.
3. Metallic purple body.

1310 ALFA ROMEO ALFASUD 89 mm 1/43 1981-1982
Details as #1310 below.
1. Silver body.
2. Red body.
3. Blue body.

1311 CAR CARRIER (not issued)

1312 PEUGEOT 505 (not issued)

1313 FORD FIESTA   83 mm   1/43   1981-1982
Details as #1313 below.
　1. Maroon body.
　2. Metallic lime green body.

1314 VOLKSWAGEN GOLF   85 mm   1/43   1981-1982
Details as #1314 below.
　1. Red body.
　2. Black body.

1315 FORD ESCORT (not issued)

1316 PEUGEOT 104 ZS   76 mm   1/43   1981-1982
Details as #1316 below, with Ecole de Pilotage decals.
　1. Silver body.
　2. White body.

1317 RENAULT 5   78 mm   1/43   1981-1982
Details as #1317 below.
　1. Yellow body.
　2. Dark blue body.

1318 RENAULT 18 (not issued)

1319 TALBOT HORIZON (not issued)

1320 PEUGEOT 305 (not issued)

THE 1300 SERIES

In April of 1982 the Cougar line that had been shared with Dinky Toys became the Solido 1300 series. Through most of the eighties, it remained Solido's main series of 1/43 scale cars.

Some of its models appeared in numerous versions, and I cannot guarantee that I have listed all of them here.

1301 CITROEN 2CV6   88 mm   1/43   Cougar>4/1982-1987
Sedan with cast body, plastic windows, closed or opened top, gray interior, gray or black chassis-grille, fast wheels, various decals.
　1. Yellow body and top, snail decals.
　2. Yellow body and top, eagle and star decals.
　3. Blue body, yellow top, no decals.

1302 CITROEN VISA   85 mm   1/43   Cougar>4/1982-1987
Sedan with cast body, plastic windows and lights, black chassis, interior and grille, fast wheels, various decals or tampo-prints.
　1. White body, #11, red-white-light blue stripes.
　2. White body, #1, same decals as type 1?.
　3. White body, #45, many more red-white-dark blue stripes.
　4. White body, orange-black-white Citroen and Total logo.
　5. White body, camel and oasis decals.
　6. White body, yellow stripe, Visa Trophee logo.
　7. Metallic green body.
　8. Light blue body, camel and oasis decals.
　9. Turquoise blue body, camel and oasis labels.
　10. Black body, camel and oasis labels.
　11. White body, Air Industrie Systemes logo. Promo.
　12. White body, Zippo log. Promo.

1303 FIAT RITMO 89 mm 1/43 Cougar>4/1982-1987

Hatchback with cast body, plastic windows and lights, black chassis and interior, spoked or fast wheels, various decals.

1. White body, red trim and Zippo logo.
2. Yellow body, #2 and orange-white-green stripes.
3. Yellow body, #3 and Alitalia logo.
4. Metallic light green body, #37, Fiat emblem and stripes.
5. Metallic green body, #37, Fiat emblem and stripes.
6. Metallic gray-green body, #37, Fiat emblem and stripes.
7. Light blue body, #37, Fiat emblem and stripes.
8. Dark blue body, white Medecin de Nuit logo.

1304 BMW 530 104 MM 1/43 Cougar>4/1982-1986

Sedan with cast body, plastic windows, black chassis, grille and interior, patterned or fast wheels, decals or tampo-print logo.

1. Silver body, #5 and Rallye du Maroc decals.
2. Silver body, #37, BP and stripes.
3. White body, #37, BP and stripes.
4. White body, #5, Fila and stripes.
5. White body, #18 and Lombard logo.
6. White body, Politi 4297 logo (Danish police car).

1305 ALFA ROMEO ALFETTA GTV 93 mm 1/43 Cougar>4/1982-1986

Fastback with cast body, black chassis, plastic windows, black interior and grille, patterned or fast wheels, various decals.

1. Gold body.
2. Silver body, #5, unknown logo.
3. Silver body, #7 and Alfa Romeo emblem.
4. Silver body, #55 and four-leaf clover.
5. Red body, #55 and four-leaf clover.
6. Red body, #27 and Agip logo.
7. Black body, #55 and four-leaf clover.
8. White body, Bendix Allied 3 logo. Promo.

1306 PEUGEOT 504 103 mm 1/43 Cougar>4/1982-1985

Sedan with cast body, plastic windows and lights, black chassis, interior and grille, silver bumpers, fast wheels, various decals or tampo-prints.

1. Gold body, white Taxi decals.
2. White body, blue * and *Ambulance* decals.
3. Yellow body, Heller Solido and 1er Janiver 1981 logo.
4. Metallic blue body, white Taxi decals.
5. Dark blue body, white Taxi labels.

1307 TALBOT TAGORA 106 mm 1/43 Cougar>4/1982-1985

Sedan with cast body, matching opening front doors, plastic windows and lights, black chassis and interior, patterned or fast wheels, +/- various decals.

1. White body.
2. White body, * and *Ambulance* decals.
3. Yellow body, Dinin batteries decals.
4. Metallic dark blue body.
5. Metallic gray body.

1308 RENAULT FUEGO 99 mm 1/43 6/1982-1987

Fastback with cast body, plastic windows and lights, black chassis-interior and grille, patterned (or fast?) wheels, various decals or tampo-prints.

1. White body, #5, blue trim and Cibie logo.
2. Dark blue body, Gendarmerie logo.
3. Gray or silver gray body? Mine may be stripped.
4. Black body, #12, cream trim and Pirelli logo.

1309 RENAULT 14 92 mm 1/43 Cougar>4/1982-1985

Sedan with cast body, matching opening front doors, plastic windows and

lights, black chassis-interior, fast wheels.
1. Gold body.
2. Light green body, rabbit decals.
3. Green body.
4. Green body, rabbit decals.
5. Orange-brown body.

1310 ALFA ROMEO ALFASUD 89 mm 1/43 Cougar>4/1982-1987

Hatchback with cast body, plastic windows, black chassis-interior and grille, patterned or fast wheels, +/- decals or tampo-prints.
1. Red body.
2. Red body, snail decals.
3. Red body, #2, green trim and other decals.
4. Red body, #56, stripes and Candy appliances logo.
5. Red body, #7 and Alfa Romeo emblem decals.
6. Blue body, #7 and Alfa Romeo emblem decals.
7. Metallic red-brown body.
8. Brown body, #7 and Alfa Romeo emblem decals.

1311 PEUGEOT 504 ? mm 1/43 1981

Apparently shown only at a 1981 toy show.

1312 PEUGEOT 505 105 mm 1/43 4/1982-1987

Sedan with cast body, matching opening front doors, plastic windows, black chassis-interior and grille-bumper, fast wheels.
1. White body, #5 and Rallye du Maroc decals.
2. Cream body, #5 and Rallye du Maroc decals.
3. Cream or off-white body.
4. Yellow body.
5. Metallic charcoal gray body.
6. White body, Police logo.
7. White body, Taxa logo. Danish issue.

1313 FORD FIESTA 83 mm 1/43 Cougar>4/1982-1985

Hatchback with cast body, matching opening doors, plastic windows and lights, black chassis, interior and bumpers, fast wheels.
1. Gold body, hand and ball on roof, figure on hood.
2. Red body, cat decals.
3. Metallic lime green body.
4. Metallic green body.

1314 VOLKSWAGEN GOLF 85 mm 1/43 Cougar>4/1982-1984

Hatchback with cast body, matching opening front doors, plastic windows, black chassis, silver gray interior and bumpers, fast wheels.
1. Red body.
2. Black body.

1315 FORD ESCORT GL 91 mm 1/43 6/1982-1986

Hatchback with cast body, matching opening doors, plastic windows and lights, black chassis, interior and bumpers, fast wheels.
1. Silver body.
2. White body.
3. Red body.
4. Metallic gray? body, #4, blue and yellow trim.
5. Black body, blue Pioneer logo, white emblem on roof.

1316 PEUGEOT 104 ZS 76 mm 1/43 Cougar>4/1982-1985

Two-door with cast body, plastic windows and lights, black grille and bumpers, black interior, fast wheels.
1. Red body, rabbit decals.
2. Red body, Banania logo decals.

1317 RENAULT 5 TL 78 mm 1/43 Cougar>4/1982-1983

Hatchback with cast body, matching opening doors, plastic windows and lights, black chassis and bumpers, gray interior, fast wheels.

1. Green body.
2. Green body, snail decals.

1318   RENAULT 18   100 mm   1/43   4/1982-1987
Sedan with cast body, matching opening front doors, plastic windows and lights, black chassis, interior and grille, silver bumpers, fast wheels.
1. Red body, "Sapeur Pompier" decals.
2. Metallic silver green body.
3. Metallic olive green body.

1319   TALBOT HORIZON   92 mm   1/43   6/1982-1985
Hatchback with cast body, matching opening front doors, plastic windows, black chassis, interior and grille, silver lights and bumpers, fast wheels.
1. Red fire car?
2. Blue body, police car.
3. Tan body.

1320   PEUGEOT 305   97 mm   1/43   6/1982-1985
Sedan with cast body, matching opening front doors, plastic windows and lights, black chassis, interior and bumpers, fast wheels.
1. White body, red and blue Police logo.
2. Light blue body, wind surf logo.
3. Metallic gray body.

1321   RENAULT 5 TURBO   82 mm   1/43   9/1982-1985
Hatchback with cast body, plastic windows and lights, usually black chassis and interior, spoked, patterned or fast wheels.
1. White body, #7 and Rallye du Var St. Maxime logo.
2. Red body, exploding star logo.
3. Metallic red body, #1, big roof spiral and Europcar logo.
4. Orange body, #1, big roof spiral and Europcar logo.
5. Orange body, #14, small roof spiral and Europcar logo.
6. Orange body, #5 and stripe decals.

1322   FIRE JEEP   81 mm   1/43   9/1982-1985
Jeep with cast body, matching windshield frame, plastic windshield, black chassis, fast wheels, spare wheel.
1. Red body, Sapeur Pompier logo.
2. Yellow body, black top, no logo.
3. Yellow body, red top, Protection Technique logo. Promo.

1323   PORSCHE 934 TURBO   100 mm   1/43   1983-1987
Sports coupe with cast body, plastic windows, black chassis, interior and airfoil, fast wheels.
1. Silver body, #5 and white-black trim.
2. White body, #61 and Shell logo.
3. White body, #69 and unknown logo.
4. Yellow body, #7, Opal and Lubrifiant logo.
5. Metallic light blue body, Club Porsche de France logo.

1324   PORSCHE 924   96 mm   1/43   1983-1985
Sports coupe with cast body, matching opening doors, plastic windows, black chassis and interior, fast wheels.
1. Metallic light blue body, Club Porsche de France logo.
2. Metallic light brown body.
3. Metallic charcoal gray body?

1325   RENAULT 4L VAN   89 mm   1/43   1983-1987
Minivan with cast body, plastic windows, white chassis, black interior and grille, patterned or fast wheels, decals or tampo-prints.
1. Red body, white Sapeurs Pompiers decals.
2. Red body, Ville de Paris decals.
3. Light blue body, yellow-on-red Solido decals.
4. Yellow body, blue windows, dome light and Flyco logo. Used in #7035 airport set.

1326 MATRA RANCHO (not sold with this number: see 2004)

1327 LANCIA RALLY 89 mm 1/43 6/1983-1987
Sports coupe with cast body, matching opening doors, plastic windows, black chassis-interior and grille, patterned or fast wheels, tampo-print logo.
1. White body, #10, stripes and Rally logo.
2. White body, Texaco logo.
3. Red body, no logo.
4. Red body, Motul 7 logo.
5. Blue body, stripes, Pirelli and Total logo.

1328 AUDI QUATTRO 102 mm 1/43 11/1983-1987
Two-door with cast body, matching opening doors, plastic windows, black chassis-interior and grille, silver lights, patterned wheels, +/- tampo-print logo.
1. Silver body.
2. White body, #8, Audi emblems, yellow and blue trim.
3. Metallic light gray body.
4. Metallic gray body.
5. Metallic gray body, Fimm-Montelimar logo. Promo.
6. Metallic gray body, Hamon Vag Dreux logo. Promo.

1329 BMW M1 104 mm 1/43 1983-1986
Sports coupe with cast body, plastic windows, black chassis-interior and louvers, fast wheels, tampo-print logo.
1. White body, #18, Lombard and Dunlop logo.
2. White body, red-blue-black trim and Le Mans logo.
3. Pale blue body, #4, yellow and blue trim.

1330 RENAULT 4L VAN 89 mm 1/43 1983-1986
Same model as #1325.
1. White body, Blommé Automation logo. Promo.

2. White body, Industrielle Electrique du Sud-Est logo. Promo.
3. White body, Electronique Generale logo. Promo.
4. White body, Securite vol Incendie Telesecurite logo. Promo.
5. White body, Loctite logo. Promo.
6. White body, Le Continent logo. Promo.
7. White body, Renault Service logo. Promo.
8. White body, Nadella Roulements logo. Promo.
9. White body, Sd Ets Deshais SA logo. Promo.
10. White body, Présidentielles 81 logo (4 versions). Promo.
11. White body, Ets Pignatta S.A. logo. Promo.
12. White body, Mr. Bricolage logo. Promo.
13. White body, Proto 9e Anniversaire logo. Promo.
14. White body, Auto Modelisme 94 logo. Promo.
15. White body, E.C.S. Transports Urgents logo. Promo.
16. White and black body, 10e Bourse de Nancy-Total logo. Promo.
17. Cream body, Rotary Club d'Issoire logo. Promo.
18. Cream body, Chasseur d'Images logo.
19. Pink body, C.C.A.M. 1978-1988 logo. Promo.
20. Yellow body, Les Electriciens de France logo. Promo.
21. Yellow body, La Telephonie Appliquee logo. Promo.
22. Yellow body, Division Industrie logo. Promo.
23. Yellow body, La Poste logo. Promo.
24. Yellow body, Depannage Entretien Maintenance logo. Promo.
25. Yellow body, Autominiateur logo. Promo.
26. Orange logo, Danish postal logo. Danish issue.
27. Light green body, Gai Jouet logo. Promo.
28. Light blue body, Telecom logo. Promo.
29. Blue logo, Lyonnaise des Eaux logo. Promo.
30. Blue body, Mercure Interim logo. Promo.
31. Purplish-blue body, Chainelec logo, many types. Promo.
32. Tan body, red Cellier des Dauphins logo. Regular issue.
33. Black and gold body, Belgom logo. Promo.

1331 JEEP RALLY 81 mm 1/43 7/1984-1987
Jeep with cast body, matching folding windshield, black plastic roll cage and chassis, fast wheels.
   1. Green body, Auto Loisirs 29 logo.
   2. Black and yellow body, #5 on multicolored square.

1332 PORSCHE 935 TURBO 108 mm 1/43 1984-1987
Sports coupe with cast body, plastic windows, black chassis-interior-grille, fast wheels, tampo-print logo.
   1. White body, #1 and blue trim.
   2. White body, RC 12 logo.
   3. Yellow body, #16, J David and black trim.

1333 ALPINE RENAULT A442B 110 mm 1/43 1984-1986
Sports-racing car with cast body, plastic windshield and lights, yellow air scoop, black chassis-interior, fast wheels, #2, Elf and black and white trim tampo-print.
   1. Yellow body.

1334 PORSCHE 936 108 mm 1/43 1984-1986
Sports-racing car with cast body, plastic lights, black chassis-interior, white air scoop and mirrors, fast wheels, #9, Boss and white trim tampo-print.
   1. Black body.

1335 PEUGEOT 504 V6 99 mm 1/43 1984-1985
Coupe with cast body, plastic windows and lights, black chassis-interior and bumpers, fast wheels, white Banzai and trim, orange-yellow-green design on roof, both tampo-prints.
   1. Orange body.

1336 PORSCHE 928 102 mm 1/43 1984-1987
Sports coupe with cast body, matching opening doors, plastic windows and lights, black chassis-interior, fast wheels.
   1. Silver body.
   2. Dark green body.
   3. Metallic light gray body.

1337 MERCEDES-BENZ 190 103 mm 1/43 7/1984-1987
Sedan with cast body, matching opening front doors, plastic windows and lights, black chassis-interior, silver grille, fast wheels or patterned wheels with red hubs.
   1. White body, green hubs.
   2. Red body, red or gold hubs.
   3. Dark green body, green hubs.
   4. Metallic gray body, green hubs.
   5. White body, green Taza logo: Danish issue.

1338 CHEVROLET CAMARO 110 mm 1/43 12/1984-1987
Sports coupe with cast body, matching opening doors, plastic windows, red interior, black chassis, patterned wheels, #1, red-white-blue trim and Ron Bailey tampo-print.
   1. White body.
   2. Metallic blue body, no logo?
   3. Black body, no logo?

1339 RENAULT 25 106 mm 1/43 5/1985-1987
Sedan with cast body, matching opening front doors, plastic windows and lights, tan or olive interior, black chassis, patterned or fast wheels.
   1. Metallic dark blue body, tan interior.
   2. Metallic dark blue body, olive interior.

3. Metallic lilac body.
4. White body, Air Industrie Systemes logo. Promo.
5. White body, Elections Présidentielles logo (5 versions).
6. Metallic blue body, Fimm-Montelimar logo. Promo.

1340   FORD SIERRA XR4i   101 mm   1/43   7/1985-1987

Two-door with cast body, plastic windows and lights, light orange interior, black chassis-lower body, patterned wheels.
1. Red body.
2. Maroon body.
3. Metallic blue body.
4. Purple body.
5. White body, Air Industrie Systemes logo. Promo.
6. Maroon body, Fimm-Montelimar logo. Promo.

1341   NISSAN PRAIRIE   93 mm   1/43   2/1986-1987

Wagon with cast body, matching opening rear hatch, tinted plastic windows, gray interior and grille-bumper, black chassis, patterned hubs.
1. Red body and hubs.
2. Metallic silver green body.

1342-1346   VISA 4X4 TELÉ-UNION   __ mm   1/43   1985-1986

Special issues: 1342 A2, 1343 RTL, 1344 Monte Carlo, 1345 SSR, 1346 Canada. No other data.

1347   PEUGEOT 205 TURBO 16   (not issued)

1348   PORSCHE 944   97 mm   1/43   3/1986-1987

Sports coupe with cast body, matching opening doors, plastic windows, tan interior, black chassis, patterned hubs.
1. Metallic light gray body.

2. Metallic dark gray body.
3. Metallic dark gray body, Fimm-Montelimar logo. Promo.

1349   PEUGEOT 205 GTI   85 mm   1/43   1/1986-1987

Hatchback with cast body, matching opening doors, tinted plastic windows and lights, tan interior, black chassis and bumpers, patterned or fast wheels.
1. White body, blue * signs.
2. Red body.
3. Red body, blue dome light, Sapeurs Pompiers logo.
4. Blue body, white Gendarmerie lettering.
5. Metallic gray body.
6. White body, Get 1, Get 2, Get 3 or Get 4 logo. Promo.
7. White body, Air Industrie Systemes logo. Promo.
8. White body, M. Bricolage logo. Promo.
9. White body, Jeanne Hachette Beauvais logo. Promo.
10. Pink body, C.C.A.M. 1978-1988 logo. Promo.
11. Yellow body, Bourse d'Echange logo. Promo.
12. Blue body, EDF-GDF logo. Promo.
13. Blue body, Electricité de France logo. Promo.

1350   FORD ESCORT RS TURBO   93 mm   1/43   3/1986-1987

Two-door with cast body, matching opening doors, plastic windows and lights, black chassis-interior and bumpers, patterned wheels.
1. White body, #2 and Metal 5 logo.
2. Yellow body.
3. Yellow body, #59 and Motul logo.
4. White body, Bendix Allied 5 logo. Promo.

1351   PEUGEOT 205 GTI   85 mm   1/43   1986-1987

Hatchback with cast body, tinted plastic windows and lights, red interior, black chassis, patterned wheels.
1. White body, #7, stripes, Newman and Monte Carlo Rally logo.

2. White body, Bendix Allied 9 logo. Promo.
3. White body, CFUVM-IBM logo. Promo.

1352 MERCEDES-BENZ 190 100 mm 1/43 1986-1987
Sedan with cast body, plastic windows and lights, black chassis-interior and bumpers, silver grille, patterned wheels.
1. White body.
2. Metallic charcoal gray body.
3. White body, Bendix Allied 7 logo. Promo.
4. Metallic charcoal gray body, Fimm-Montelimar logo. Promo.

1353 RENAULT MAXI 5 TURBO 86 mm 1/43 4/1986-1987
Two-door with cast body, plastic windows and lights, yellow chassis and seats, white dash and rear interior, patterned wheels, #4, Opal, blue, yellow and black trim.
1. Red body.
2. Black body.
3. Black body, Coches Llenos de Vida logo. Promo.
4. White body, Bronze Acior logo. Promo.
5. White body, Bendix Allied 1 logo. Promo.

1354 ALFA ROMEO ALFETTA GTV 95 mm 1/43 1/1987
Fastback with cast body, plastic windows, tan interior, black chassis and grille, patterned wheels.
1. Black body.

1355 BMW M1 100 mm 1/43 1987
Sports coupe with cast body, plastic windows, black chassis and interior, white dash and louvers, fast wheels.
1. Red body.

1356 PORSCHE 959 COUPE (announced 1987, not issued)

1357 RENAULT SUPER 5 77 mm 1/43 1987
Hatchback with cast body, plastic windows and lights, tan interior, black chassis and bumpers, wheels with red hubs.
1. Red body, tan or black grille.
2. Gold body and chassis, Renault Flins 1987 logo. Promo.

1358 VOLKSWAGEN GOLF GTI 86 mm 1/43 7/1987
Hatchback with cast body, plastic windows, black chassis, interior and grille, patterned wheels.
1. Red body.
2. Dark red body.
3. Black body.

1359 CITROEN 2CV6 88 mm 1/43 1/1987
Open-top sedan with cast body, red plastic interior, black chassis-grille, wheels with red hubs, #18 and Echappement logo.
1. Yellow body.
2. White body, Automobiles Anciennes de Troyes logo. Promo.
3. Yellow body, Relais Culturel de Thann logo. Promo.
4. Olive green body, Gai Jouet logo. Promo.
5. Cream or gray body?, 2CV Club de France logo. Promo.
6. Gray body.

1365 CITROEN CX 2400 AMBULANCE 113 mm 1/43 1987
Ambulance with cast body, matching opening hatch, black chassis, blue plastic windows, headlights and roof light, tan interior, gray bumpers, fast wheels, blue * and stripes.
1. White body.
2. Red body, Pompiers logo.
3. Metallic red body.
4. Metallic gray body.

THE 1400 SERIES

Numbers 1401 to 1406 were used for the models made for sale as Dinky Toys, paralleling the 1300 numbers of the first six Cougar versions of the same models.

HI-FI AND TO DAY SERIES

This series of resurrections, soon joined by new models, began early in 1988 as the Hi-Fi Series and changed its name to To Day in 1991. Though a few models have been dropped, the series is still in existence. There may be more color variations than are listed here. The dates are those on the bases of the models.

1501 JAGUAR XJ 12 110 mm 1/43 2/1988-
Sedan with cast body, matching opening front doors, black chassis, plastic windows, tan interior, silver grille, bumpers and hubs, tires, suspension.
1. Maroon body.
2. Green body.
3. Metallic dark green body.

1502 PORSCHE 944 96 mm 1/43 1/1988-
Sports coupe with cast body, matching opening doors, black chassis, plastic windows and lights, black interior, wire hubs, tires, suspension.
1. White body.
2. Red body.
3. Blue body.

1503 ALPINE 4310 96 mm 1/43 1/1988-1991
Sports coupe with cast body, matching opening doors, plastic windows and opening hatch, black chassis and interior, gray hubs, tires, suspension.
1. Metallic blue body.
2. Metallic dark blue body.
3. Red body, Club Alpine 63 logo. Promo.

1504 RENAULT 25 106 mm 1/43 3/1988-1989
Sedan with cast body, matching opening front doors, plastic windows and lights, tan interior, black chassis and bumpers, wheels with patterned hubs, suspension.
1. Maroon body.
2. Metallic lilac body.
3. Metallic gray body.
4. Metallic brownish-gray body.
5. White body, Europe 1 logo.
6. Yellow body, Porte ouverte Solido 1988 logo. Club model.

1505 PORSCHE 928 102 mm 1/43 3/1988-1991
Sports coupe with cast body, matching opening doors, black chassis, plastic windows and lights, black interior, silver hubs, tires, suspension.
1. Yellow body.
2. Blue body.
3. Metallic dark blue body.
4. Metallic brown body.

1506 MERCEDES-BENZ 190 101 mm 1/43 1/1988-1989
Sedan with cast body, matching opening front doors, plastic windows and lights, black chassis-interior, silver grille, wheels with silver hubs, suspension.
1. Pale gold body.
2. Metallic gray body.
3. Metallic brownish-gray body.

1507 CHEVROLET CAMARO Z28 110 mm 1/43 3/1988-
Sports coupe with cast body, matching opening doors, black chassis, plastic windows, black interior, star hubs, tires, suspension.
1. Dark red body.
2. Metallic green body.
3. Metallic blue body.
4. Black body.

1508 PEUGEOT 205 GTI 85 mm 1/43 1988-
Hatchback with cast body, matching opening doors, plastic windows and lights, black chassis, interior and bumpers, wheels with silver hubs.
1. Silver body.
2. Red body.
3. Black body.
4. White body, Radio France Toulouse logo. Promo.
5. White body, Europe 1 logo.
6. Pearl white body, Darl'mat logo. Promo.
7. Red body, La Mule d'Or logo. Promo.

1509 CHEVROLET CAMARO RACING 110 mm 1/43 3/1988-1989
Same model as #1507, with #28 and black grille decals.
1. Red body.

1510 MERCEDES-BENZ 190 16S 100 mm 1/43 1988-
Same basic model as #1506, plus blue-black grille and airfoil, #50 and yellow Monroe decal.
1. Blue-black body.

1511 ROLLS-ROYCE CORNICHE 123 mm 1/43 6/1988-
Open convertible with cast body, black chassis, plastic windows, white folded top, light tan interior, silver grille, lights and bumpers, wheels with silver hubs and whitewalls, suspension.

1. Metallic dark green body.
2. Metallic light blue body.
3. Metallic brown body.

1512 BENTLEY CONTINENTAL 123 mm 1/43 6/1988-
Convertible with cast body, black chassis, plastic windows, white top, tan interior, silver grille, lights and bumpers, wheels with silver hubs and whitewalls, suspension.
1. Red body.

1513 CHEVROLET CORVETTE COUPE 103 mm 1/43 4/1989-
Sports coupe with cast body, black chassis, plastic windows and opening hatch, black interior, spoked hubs, tires, suspension.
1. Metallic dark red body.

1514 CHEVROLET CORVETTE CONVERTIBLE 103 mm 1/43 4/1989-
Open convertible with cast body, black chassis, plastic windshield, black windshield frame and interior, spoked hubs, tires, suspension.
1. Yellow body.

1515 FERRARI BB 99 mm 1/43 1/1989-
Sports coupe with cast body, matching opening rear hood, black chassis, plastic windows, black interior, lights and parts, gray motor and exhaust pipes, star hubs, tires.
1. Red body.

1516 PEUGEOT 605 SV 24 110 mm 1/43 5/1990-
Sedan with cast body, matching opening front doors, black chassis, tinted plastic windows and lights, black interior and bumpers, patterned hubs, tires, suspension.
1. Metallic magenta body, black stripes.

1517 MERCEDES-BENZ 600SL CABRIOLET 102 mm 1/43 1991-
Open convertible with cast body, plastic windshield and lights, reddish interior and folded top, black chassis and grille, wheels with silver hubs, suspension.
   1. Metallic dark red body.

1518 MERCEDES-BENZ 600SL HARDTOP 102 mm 1/43 1991-
Hardtop with cast body, plastic windows and lights, tan interior, black chassis and grille, wheels with silver hubs, suspension.
   1. Black body, silver lower panels.

1519 RENAULT CLIO 84 mm 1/43 1991-
Hatchback with cast body, matching opening doors, black chassis, plastic windows and lights, black interior and wiper, wheels with silver hubs, suspension.
   1. Red body, black stripes.

1520 RENAULT CLIO 84 mm 1/43 1991-
Same model as #1519, with gray-brown interior.
   1. Metallic blue-black body, black stripes.

1521 BMW SERIES 3 102 mm 1/43 1991-
Sedan with cast body, matching opening hood, dark gray chassis with data, plastic windows and lights, gray-brown interior, black engine, silver exhaust pipe, wheels with silver hubs, suspension.
   1. Metallic dark blue body.

1522 RENAULT ESPACE 100 mm 1/43 1991-
Minibus with cast body, matching opening hatch, plastic windows and lights, black chassis and interior, wheels with silver hubs, suspension.
   1. White body, orange and blue Expo 92 labels.
   2. Metallic blue-black body.

1523 CITROEN ZX 92 mm 1/43 1991-
Sedan with cast body, matching chassis and opening front doors, plastic windows and lights, dark gray interior and bumpers, silver muffler, wheels with silver hubs, suspension.
   1. Metallic gray body.

1524 CITROEN ZX SPORT 92 mm 1/43 1991-
Same model as #1523, with black rear awning, different hubs.
   1. Red body, chassis and doors.

1525 PORSCHE 928 GT 103 mm 1/43 1991-
Sports coupe with cast body, matching opening doors, black chassis, plastic windows and lights, black interior and airfoil, wheels with silver hubs, suspension.
   1. Metallic brown body.

1526 RENAULT CLIO 16S 84 mm 1/43 1992-
Same model as #1519, with gray-brown interior, #91 and other decals, sheet of decals enclosed.
   1. Blue body, chassis and doors.

1527 LAMBORGHINI DIABLO 1992-

1528 RENAULT 1992 or 1993?

SOLIDO 2 SERIES

This was a series of limited editions, usually based on previous models of racing and rally cars and supplied with decals to represent actual competition cars. In 1982 the series was turned over to the TOP 43 firm, which issued the last few models and expanded the series, using a different numbering system.

1700 PORSCHE 908 LE MANS 1980   mm 1/43 12/1980-1982
Same model as #86, with #6 and Martini decals.
  1. White body.

1701 FIAT ABARTH 131 SPA 1979 95 mm 1/43 12/1980-1982
Same model as #54, with #1 and Carling Black Label decals.
  1. Black body.

1702 PORSCHE 935 LE MANS 1979 108 mm 1/43 12/1980-1982
Same model as #1032, with #39 and Kores decals.
  1. Black body.

1703 PORSCHE 934 LE MANS 1980 100 mm 1/43 12/1980-1982
Same model as #68, with #16 and Denver decals.
  1. White body.

1704 FIAT ABARTH 131 PORTUGAL 95 mm 1/43 4/1981-1982
Same model as #54, with #5 and Fiat decals.
  1. White body.

1705 LANCIA STRATOS TOUR DE FRANCE 84 mm 1/43 4/1981-1982
Same model as #73, with #3 and Le Point decals.
  1. Black body.

1706 PORSCHE 935 JOHN PAUL 108 mm 1/43 6/1982
Same model as #1032, issued by Top 43.
  1. Light blue body.

1707 BMW M1 BASF   mm 1/43 12/1981-1982
Same model as #1329, with #76 and BASF decals.
  1. Red body.

1708 BMW 530 SPA 1978 104 mm 1/43 4/1981-1982
Same model as #89, with #3 and Kinley decals.
  1. White body, yellow chassis.

1709 PORSCHE 935 KREMER 108 mm 1/43 9/1981-1982
Same model as #1032, with #2 and Vaillant decals.
  1. Green body.

1710 BMW 530   (not issued)

1711 RENAULT 5 TURBO MONACO 82 mm 1/43 12/1981-1982
Same model as #1023, issued by Top 43.
  1. Yellow body, white chassis.

1712 PORSCHE 934 WHITTINGTON 100 mm 1/43 1/1982
Same model as #68, with #93 and Whittington decals. Issued by Top 43.
  1. Yellow body.

YESTERDAY SERIES

Here we have another series of reissues. It began in 1992 and includes cars with a sporting or sports-related atmosphere, some of them with racing or rally decals.

1801   MASERATI INDY 1970   110 mm   1/43   1992-
Same model as #185, with opening doors and rear hatch, black chassis-interior, wheels with silver hubs.
   1. Metallic silver gray body and doors.

1802   FERRARI BB 1976   100 mm   1/43   1992-
Same model as #1515, with black cast chassis and plastic interior and parts, wheels with star hubs.
   1. Red body and rear hood.

1803   ALPINE A110 1970   89 mm   1/43   1992-
Same model as #181, with opening doors, black cast chassis and plastic interior, spoked hubs, tires.
   1. Metallic blue body and doors.

1804   ALPINE A110 MONTE CARLO 1973   89 mm   1/43   1992-
Same model as #1803, with #18 and Rallye Monte Carlo decals.
   1. Metallic blue body and doors.

1805   OPEL GT 1968   95 mm   1/43   1992-
Same model as #171, with opening doors, black plastic chassis and interior, silver hubs, tires.
   1. Yellow body and doors.

1806   JAGUAR XJ 12 1978   110 mm   1/43   1992-
Same model as #1501, with opening front doors, black cast chassis, tan plastic interior, silver hubs, tires.
   1. Metallic brown body and doors.

The following models were probably not issued:
1807   CITROEN SM 1970
1808   PORSCHE CARRERA 1973
1809   LANCIA STRATOS 1978
1810   FERRARI 365 GTB4 1972
1811   FERRARI 365 GTB4 RACING
1812   BMW M1 1979

THE 2000 SERIES

The 2000 series grew out of the old 300-3300 series and was, I suspect, renumbered to indicate lower price than the bigger 3000 series commercial and emergency vehicles. The series was instituted in 1983, and two of its members lasted until 1988. Each of the ten models had been issued under at least one other number before, and half of them were given new numbers later, when the 2100 series was introduced.

2000   HOTCHKISS FIRE TRUCK   111 mm   #362>#3362>1983-1985>#2100

Same model as #362.
1. Red body.

2001   SAVIEM FIRE TRUCK   112 mm   1/50   #3303>1983-1985>#2101

Same model as #3303, with fast wheels, big Sapeur Pompier decals.
1. Red cab and body.

2002   CITROEN C35 VAN   98 mm   1/50   #368>#3368>1983-1986>#2116

Same model as #368 fire ambulance, but also issued with black Milleville decals, white roof and interior, fast wheels.
1. Red body, white roof (fire ambulance).
2. Yellow body, white roof, Milleville decals.

2003   PEUGEOT J7 POLICE BUS   91 mm   1/50   #372>#3372>1983-1986

Same model as #372, with fast wheels.
1. Blue body, white roof.

2004   TALBOT MATRA RANCHO   103 mm   1/43   #1062>1983-1986

Same model as #1062, with fast wheels.
1. Green body.

2005   LAND-ROVER 109   109 mm   1/43   #66>#1066>1983-1986

Same model as #66, with Transaharienne, map and camel decals, fast wheels. Promo has luggage rack on roof.
1. Tan body, white roof.
2. Tan body, black roof and rack, Camel Trophy logo. Promo.
3. Tan body, Operation Polio Plus and Rotary International or Rotary Club Evreux logo. Promo.

2006   SAVIEM WRECKER   115 mm   1/50   #366>#3300>1983-1988

Same model as #366, with SOS depannage decals, fast wheels.
1. Yellow cab, blue-black rear body.

2007   DODGE TRUCK   120 mm   1/50   1983-1988

Same model as #242 or #391, but in civilian form, with no cover, fast wheels.
1. Yellow body and chassis.

2008   BERLIET FIRE TRUCK   116 mm   1/55   #375>#3375>1984-1986> #2106

Same model as #375, with fast wheels.
1. Red body, white roof.

2009   IVECO DUMP TRUCK   135 mm   1/60   #374>#3374>1984-1986>#2106

Same model as #374, with fast wheels.
1. Red cab, yellow tipper, black chassis.

2100 SERIES

This series began in 1985, and I don't know why it replaced the 2000 series, from which five of its first models were taken. It consists exclusively of emergency and other public service vehicles, most of which, other than its newest members, were moved to it from other series.

2100 HOTCHKISS FIRE TRUCK 111 mm #362>#3362>#2001>1985-1987
Same model as #2000.
1. Red body.
2. Red body, Paris HQ Fire Dept. logo.

2101 SAVIEM FIRE TRUCK 111 mm 1/50 #3303>#2001>1985-
Same model as #2001.
1. Red cab and body.

2102 SAVIEM FIRE WRECKER 110 mm 1/50 #366B>#3301>1985-1987 Same model as #3301.
1. Red cab and body.

2103 BERLIET FOREST FIRE TRUCK 145 mm 1/50 #354>#3354>1985-1988
Same model as #3354.
1. Red body.

2104 FIRE LAND-ROVER 106 mm 1/43 #1034>1985-1987
Same model as #1034. (Promos listed under #2005.)
1. Red body, white roof.

2105 DODGE FIRE TRUCK 119 mm 1/50 #391>#3391>1985-1987
Same model as #3391, with Ville de Paris decal.
1. Red body and chassis, tan cover.

2106 BERLIET FIRE TRUCK 116 mm 1/55 #375>#3375>#2008>1985-Same model as #2008.
1. Red body.

2107 BERLIET FIRE ENGINE 123 mm #350>#3350>1985-1987
Same model as #350.
1. Red body.

2108 SIMCA-UNIC SNOWPLOW TRUCK 142 mm #359>#3359>1985-1987
Same model as #359.
1. Yellow body, orange cover.
2. Blue body, silver gray cover.
3. Red body, Mont Blanc logo.
4. Orange body, D.D.E. logo.

2109 IVECO DUMP TRUCK 135 mm 1/60 #374>#3374>#2009>1985-1987
Same model as #2009.
1. Red cab, yellow tipper, black chassis.

2110 VOLVO ROAD ROLLER 102 mm 1/55 #367>1985-1987
Roller with cast body, cab, chassis, unpainted roller and parts, yellow plastic roller mount, silver gray stack, blue windows, plastic wheels with silver hubs.
1. Yellow body, cab and chassis.

2111 IVECO BUCKET TRUCK 1985-1987
No data.

2112 IVECO TELE-UNION TRUCK 1986
Special issue. No data.

2113 SIMCA-UNIC FOREST FIRE TRUCK 107 mm 1986-1987
Covered truck with cast cab and rear body-chassis, unpainted drive train, plastic windshield, silver gray cover, wheels with red hubs, S.D.I. de la Drôme decals.
1. Red cab and body.

2114 FIRE JEEP AND PUMP TRAILER 131 mm 1/43 3304>1986-1987 Same model as #3304.
1. Red Jeep body and trailer.

2115 PEUGEOT J7 FIRE BUS 91 mm 1/50 #372>#3372>1986-1987
Same model as #372, with Ville de Paris decals, wheels with red hubs. White ADP version is from #7035 airport set.
1. Red body, white roof.
2. White body and roof, ADP decals.

2116 CITROEN AMBULANCE VAN 100 mm 1/50 #368>#3368>2002> 1986-1987
Same model as #2002 fire ambulance, but also in white version.
1. Red body, white roof, Sapeurs Pompiers Ambulance... logo.
2. White body and roof, blue Ambulance logo, no tow hook.
3. White body, Filtoys logo. Promo.
4. White body, Compagnie Vosigienne Isolation logo. Promo.
5. Red body, white roof, white chevrons (Citroen emblem). Promo.
6. Olive green body, 3 X 34 Transport logo. Promo.
7. Red body, Dyreambulance-Falck logo. Danish issue.
8. Orange body, Danish postal logo. Danish issue.
9. White body, A-Post logo. Danish issue.
10. Blue body, Vitrines Jean Groene logo. Promo.
11. White body, Ragain logo. Promo.

2117 JEEP FIRE ENGINE 81 mm 1/43 1987-
Jeep with cast body and matching folding windshield frame, black plastic chassis and pump, red rear box, white hose reel, wheels with red hubs, white S.D.I. du Var Cogolin logo on hood.
1. Red body.

2118 CITROEN C35 FIRE VAN 100 mm 1/50 1987-1989
Same casting as #2116, with white roof rack, silver ladders, wheels with red hubs, white Sapeurs Pompiers de Paris logo.
1. Red body, white roof and opening rear doors.
2. Red body, Saint Malo 1987 - CNP logo. Promo.

2119 TALBOT MATRA RANCHO FIRE WAGON 104 mm 1/43 1987-1988
Wagon with cast body, blue plastic windows, lights, dome light and opening hatch, black chassis, interior and parts, wheels with red hubs, white Sapeurs Pompiers logo. Mine lacks grille.
1. Red body.

2120 PEUGEOT J9 VAN 91 mm 1/50 1987-1988
Van version of #2115, with cast body, black chassis, blue plastic windows and dome light, white roof and opening rear doors, gray siren, wheels with red or silver hubs, white logo.
1. Red body, Sapeurs Pompiers logo.
2. Blue body, Gendarmerie logo.
3. Blue body, Politi logo. Danish issue.

2121 SIMCA-UNIC FIRE TRUCK 110 mm 1987-1990
Truck with cast cab and chassis, plastic windshield, red rear body with white parts, red and white hose reel, wheels with red hubs, white S.D.I. du Tarn logo.
1. Red cab and chassis.

2122 RENAULT FIRE VAN & PUMP TRAILER 145 mm 1/43 5/1988-1989

Minivan and trailer with cast bodies. tinted van windows, lights and dome light, black chassis, wheels with red hubs, white Ville de Saint-Malo logo.
1. Red bodies.

2123 DODGE FIRE TRUCK 123 mm 1/50 1988-1989

Dodge truck body like #2105, with red rear body of #2121, silver gray cab roof, wheels with red hubs, white St. Tropez logo.
1. Red body and chassis.

2124 SIMCA-UNIC FIRE WRECKER 128 mm 1988-1989

Same cab and chassis as #2121, with open rear, black plastic wrecker boom, wheels with red hubs, Sapeurs Pompiers decals.
1. Red cab and chassis.

2125 MERCEDES UNIMOG FIRE TRUCK 112 mm 1/50 5/1988-

Truck with cast cab and chassis, tinted plastic windows and dome light, red rear body, silver gray panels, black inner chassis, grille and exhaust stack, wheels with red hubs, Sapeurs Pompiers Mulhouse decals.
1. Red cab and chassis.

2126 PEUGEOT J9 FIRE AMBULANCE 91 mm 1/50 1989-

Van with cast body, black chassis, blue plastic windows and dome lights, red opening rear doors, gray interior and siren, wheels with red hubs, white Brigade de Sapeurs Pompiers Ambulance logo.
Van has rear side windows, unlike #2120.
1. Red body.

2127 MERCEDES UNIMOG FOREST FIRE TRUCK 122 mm 1/50 1989-1990

Truck with cab, chassis and parts of #2125, plus red plastic rear body with white hose reel, spare wheel, white Feuerwehr logo.

1. Red cab and chassis.

2128 DODGE FIRE VAN 97 mm 1/50 1989-

Van with cast body and chassis, plastic windows, red siren and opening rear doors, black inner chassis and interior, wheels with red hubs, white Ville de Séestat logo.
1. Red body and chassis.

2129 IVECO FIRE TRUCK 112 mm 1990-

Covered truck with cast cab, chassis and rear body, blue plastic windows and dome light, red cover, black front fenders, gray inner chassis, wheels with red hubs, white Sapeurs Pompiers logo.
1. Red cab, body and chassis.

2130 RENAULT VAN AND PUMP TRAILER 1/43 1990-

Appears to be revision of #2122.

2131 MERCEDES 407D RESCUE VAN 106 mm 1/50 9/1990-

Van with cast body, blue plastic windows and dome lights, red opening rear doors, black chassis, grille and interior, wheels with red hubs, red Feuerwehr on white side panel.
1. Red body.

2132 MERCEDES 407D FIRE VAN 106 mm 1/50 9/1990-

Same basic model as #2131, with different interior, black roof rack, silver ladders, one dome light, white Feuerwehr on hood.
1. Red body.

2133 MERCEDES UNIMOG RESCUE VAN 110 mm 1/50 1/1991-

Truck with same cast cab, chassis and plastic parts as #2127, plus red plastic rear body with blue dome light, white #4 and Notruf 112 logo.
1. Red cab and chassis.

2134 MERCEDES UNIMOG WRECKER 135 mm 1/50 1992-
Truck with same cast cab and chassis as #2133, with cast open rear body, black plastic wrecker boom, white telephone, #18 and stripes.
  1. Red cab, chassis and body.

2200 SERIES

Many of the 200 series of military vehicles were renumbered in 1980. None of them stayed in production longer than 1982, and no models were changed physically. In some cases, two color variations of the same model were given two different numbers, otherwise the new number was the old number prefixed by a 2.

2200 M-20 COMBAT CAR 100 mm #200>1980

2201 UNIC ROCKET LAUNCHER 177 mm #201>1980

2202 PATTON M47 TANK 137 mm #202>1980-1982
The olive US version.

2203 RENAULT 4X4 TRUCK 96 mm #203>1980-1982

2204 105 mm CANNON 229 mm #204>1980-1982

2205 105 mm CANNON ON WHEELS 190 mm #205>1980-1982

2206 250/0 HOWITZER 155 mm #206>1980

2207 SU 100 TANK 194 mm #208S>1980
The tan Egyptian version.

2208 SU 100 TANK 194 mm #208>1980-1981
The Russian version.

2209 AMX 30 TANK 170 mm #209>1980-1982
The olive French version.

2210   AMX 30 A-A TANK   170 mm   #209B>1980-1982
The tan Egyptian version.

2211   BERLIET TANK TRANSPORTER   #211>1980-1982

2212 BERLIET TRANSPORTER & AMX TANK #211B>1980-1982

2213   AUTO-UNION JEEP AND TRAILER   #212-213>1980

2214   PATTON M47 TANK   137 mm   #202S>1980
The tan Israeli version.

2218   PT 76 ROCKET TANK   144 mm   #218-1980

2222   TIGER I TANK   #222>1980
The gray German version.

2223   TIGER I TANK   #222B>1980
The tan/olive camouflaged version.

2226   BÜSSING ARMORED CAR   114 mm   #226>1980-1982

2227   AMX 13 VTT TANK   112 mm   #227B>1980

2228   JAGDPANTHER TANK   196 mm   #228>1980
The gray German version.

2229   JAGDPANTHER TANK   196 mm   #228B>1980
The tan/maroon camouflaged version.

2231   SHERMAN M4 TANK   124 mm   #231>1980-1981

2232   M10 TANK DESTROYER   #232>1980

2233   RENAULT R35 TANK   95 mm   #233>1980-1981

2234   SOMUA S35 TANK   107 mm   #234>1980

2236   PANTHER G TANK   175 mm   #236>1980

2237   PzKW IV TANK   142 mm   #237>1980-1982

2238   AMX ROCKET LAUNCHER   #238>1980-1982

2239   PANHARD AML H90   108 mm   #240S>1980
The tan version.

2240   PANHARD AML H90   108 mm   #240>1980
The olive version.

2241   HANOMAG HALFTRACK   120 mm   #241>1980

2242   DODGE ARMY TRUCK   120 mm   #242>1980-1982

2243   LEOPARD TANK   #243>1980

2244   M3 HALFTRACK   127 mm   #244>1980-1981

2245   KAISER-JEEP ARMY TRUCK   140 mm   #245>1980-1981
A.H.X. 10

2247   BERLIET-ALVIS   125 mm   #247>1980-1981

2248   M41 TANK DESTROYER   #248>1980

2249   AMX 13 TWO-GUN TANK   110 mm   #249>1980

2250   AMX 13 TANK WITH 90 GUN   134 mm   #250>1980

2251 SAVIEM VAB AMPHIBIAN   #251>1980-1982

2252 M7B1 PRIEST   118 mm   #252>1980-1981

2253 GENERAL LEE TANK   116 mm   #253>1980-1982

2254 AMX 10P TANK   #254>1980

2255 BERLIET AIRPORT FIRE TRUCK 140 mm #255>1980-1981

2256 US ARMY JEEP & TRAILER   #256>1980-1982

2257 SAVIEM TANKER SEMI   215 mm   #257>1980-1981

2259 CITROEN C35 ARMY AMBULANCE   98 mm   #259>1980-1981

2262 RICHIER ARMY CRANE   140 mm   #262>1980-1981

RETROSPECTIVE LE MANS SERIES

This was yet another series of reissues, composed of two sets, each containing two versions each of six models--with one exception, the pair of similar Matras--with decals to represent Le Mans competitors of past years. Each series was apparently a one-shot issue, and was sold to dealers in white cardboard boxes containing one each of the twelve models. The first set, the individual models of which were numbered 2401 to 2412, was numbered 7153 and issued in 1990. The second set, of models 2413 to 2424, was numbered 7160 and issued in 1992.

2401 DB PANHARD 1959 84 mm 1/43 1990
Same model as #112, with cast body, black chassis-interior, plastic windshield, silver hubs, tires, #50 decals.
   1. Blue body.

2402 DB PANHARD 1960 84 mm 1/43 1990
Same model as #2401 but with #48 decals.
   1. Blue body.

2403 FERRARI GTO 1962 101 mm 1/43 1990
Same model as #4506, with cast body, matching opening doors, black chassis-interior, plastic windows and lights, silver exhaust pipes, wheels with wire hubs, #19 decals.
   1. Red body.

2404 FERRARI GTO 1963 101 mm 1/43 1990
Same model as #2403, with #25 decals.
   1. Silver body.

2405 ALPINE RENAULT A220 1968 108 mm 1/43 1990
Same model as #168, with cast body, matching opening doors and rear hood, plastic windows, black chassis-interior, silver motor and hubs, tires, suspension, #30 decals.
   1. Blue body.

2406 ALPINE RENAULT A220 1968 108 mm 1/43 1990
Same model as #2405, with #27 decals.
   1. Blue body.

2407 MATRA-SIMCA 670 1972 (short) 98 mm 1/43 1990
Same model as #13, with cast body, matching opening rear hood, black chassis, plastic lights, white cowling and air scoop, silver motor, wheels with spoked hubs, #15 decals.
   1. Blue body.

2408 MATRA-SIMCA 670 1972 (long) 104 mm 1990
Same model as #14, with castings and parts as above plus silver airfoil and #14 decals.
   1. Blue body.

2409 FERRARI 365 GTB4 1973 98 mm 1/43 1990
Same model as #16, with cast body, matching opening doors, plastic windows and lights, black chassis-interior, wheels with star hubs, suspension, #40 and Thomson decals.
   1. White body.

2410 FERRARI 365 GTB4 1974 98 mm 1/43 1990
Same model as #2409, with #71 decals.
   1. Red body.

2411 PORSCHE 935T 1979 108 mm 1/43 1990
Same model as #1332, with cast body, black chassis, plastic windows, black interior, wheels with yellow hubs, #71 decals.
   1. Yellow body.

2412 PORSCHE 935T 1979 108 mm 1/43 1990
Same model as #2411, with silver hubs and #68 decals.
   1. Black body.

2413 FERRARI TCR 1957 94 mm 1/43 1992
Same model as #103, with cast body, plastic windshield, brown seats, wire hubs, #28 decals.
   1. Yellow body, blue trim.

2414 FERRARI TCR 1957 94 mm 1/43 1992
Same model as #2414, with black seats, wire hubs, #29 decals.
   1. Light blue body.

2415 FORD MARK IV 1967 102 mm 1/43 1992
Same model as #170, with cast body, black chassis and interior, plastic windows and lights, wheels with spoked hubs, #3 decals.
   1. Metallic bronze body.

2416 FORD MARK IV 1967 102 mm 1/43 1992
Same model as #2415, with #4 decals.
   1. Dark blue body.

2417 LIGIER JS3 1971 88 mm 1/43 1992
Same model as #195, with cast body and chassis, with plastic lights, light gray cowling, black seats, silver roll bar and intakes, wheels with spoked hubs, #32 decals.
   1. Yellow body, green chassis.

2418   LIGIER JS3 1971   88 mm   1/43   1992
Same model as #2417, with #24 decals.
   1. Yellow body, green chassis.

2419   ALFA ROMEO 33/3   102 mm   1/43   1992
Same model as #187, with cast body and opening rear hood, plastic windshield and lights, black chassis and interior, silver motor and roll bar, red mirror and airfoil, wheels with silver hubs, #37 decals.
   1. Red body and hood.

2420   ALFA ROMEO 33/3   102 mm   1/43   1992
Same model as #2419, with #38 decals.
   1. Red body and hood.

2421   PORSCHE 917 1971   95 mm   1/43   1992
Same model as #186, with cast body and opening rear hood, plastic windows and lights, black chassis and parts, silver motor, wheels with spoked hubs, #19 decals.
   1. Pale blue body and hood.

2422   PORSCHE 917 1971   95 mm   1/43   1992
Same model as #2421, with Martini and #22 decals.
   1. White body and hood.

2423   LOLA T280 1972   84 mm   1/43   1992
Same model as #15, with cast body and opening doors, plastic lights, black chassis and interior, silver roll bar and airfoil, wheels with spoked hubs, #8 decals.
   1. Dark yellow body and doors.

2424   LOLA T280 1973   84 mm   1/43   1992
Same model as #2423, with #61 decals.
   1. Dark yellow body and doors.

3000 SERIES

Like the 2000 series, the 3000 series came into being in 1983; it was gone by the end of 1985, with only one model surviving to enter the 3100 series that succeeded it. Six of its seven members were vehicles towing something; the last was an Iveco truck that, like the first, had been moved in from the 3300 series.

3000   CITROEN FIRE VAN & TRAILER   102 mm   1/50   #371>#3371-1983-1985>#3100
Same model as #371.
   1. Red body, white roof, brown and gray boat on red trailer.

3001   LAND ROVER SAFARI & CAGE TRAILER   199 mm   1/43   1983-1985
Same Land-Rover as #2005, with black roof rack, fast wheels, Reserve Africaine and coat of arms, pulling tan plastic flat trailer with white cage, fast wheels.
   1. Tan body and chassis, white roof.

3002   SAVIEM WRECKER AND RENAULT 5   190 mm   1/50-43   1983-1985
Same wrecker as #366, with unpainted boom, white hooks and SOS Depannage logo, and fast wheels, plus #1321 Renault with #1 and Europcar logo and big spiral on roof.
   1. White wrecker cab, red body, orange car.

3003   MATRA RANCHO & HORSE TRAILER   188 mm   1/43   1983-1985
Same Rancho as #1062, with black chassis, interior and parts, and fast wheels, pulling trailer with cast body, silver chassis, black plastic tailgate, fast wheels.
   1. Metallic dark blue Rancho, metallic golden brown trailer.

3004 DODGE FIRE TRUCK & PUMP TRAILER 169 mm 1983-1985
Open truck with cast body parts and chassis, black plastic grille, pulling cast trailer with black plastic chassis, both with fast wheels. Pump as #360, truck resembles #391.
  1. Red truck body, chassis and trailer body.

3005 SIMCA-UNIC SNOWPLOW TRUCK & TRAILER 204 mm 1983-1985>
#3600
Same truck as #359, with silver cast plow and mount, fast wheels with tire chains, pulling open two-wheel trailer with cast body, black plastic chassis, fast wheels.
  1. Yellow truck and trailer bodies, orange cover.

3007 IVECO ESSO TRUCK 113 mm #3307>1985
Covered truck with cast cab and rear body, silver chassis, tinted windows, white cover, black inner chassis, silver gray front fenders and hubs, tires, red Team Esso aseptogyl logo decals.
  1. White cab and body.

3100 SERIES

In 1985 the 3100 series was instituted, apparently as a home for the lower-priced models that had been in the 3300 series until then, for this is where the earlier members of the series came from. It has since been expanded and now consists completely of fire vehicles.

3100 CITROEN FIRE AMBULANCE AND LIFEBOAT 192 mm 1/50 #371> #3371>#3000>1985-1986
Same model as #3000, with Sauvetage en Mer Marins Pompiers, Ville de Bastia logo, wheels with red hubs.
  1. Red body, white roof, brown and gray boat.

3101 MERCEDES-BENZ LADDER TRUCK 150 mm 1/50 #361>#3361>1985-1986
Same model as #361, presumably with fast wheels.
  1. Red body.

3102 RICHIER FIRE CRANE 139 mm 1/73 #353>#3353>1985-
Same model as #353, with white cast chassis and plastic hook, fast wheels, white Sapeurs Pompiers de Paris or red-on-white Service Departementale logo.
  1. Red body and main boom, white Sapeurs Pompiers de Paris logo.
  2. Red body and main boom, red Service Departemental logo.

3103 VOLVO DUMP TRUCK 160 mm 1/74 #356>#3356>1985
Same model as #356, presumably with fast wheels.
  1. Yellow body.

3104 VOLVO WHEEL LOADER 129 mm 1/55 #367>#3367>1985
Same model as #367.
  1. Yellow body.

3105 MERCEDES-BENZ OVERHEAD SERVICE TRUCK 149/175 mm #358> #3358>1985

Same model as #3105. No other data.

3106 MACK FIRE ENGINE 138 mm 1/60 1986-

Fire truck with cast body, silver chassis, plastic windows, silver ladder and parts, white hoses and inner chassis, wheels with red hubs, control panel labels, white grid deck panels.
1. Red body, Engine 4, Brewster, Mass. decals.
2. Light orange body, same decals as type 1.
3. Yellow body, County of Orange, California decals.
4. Lime green body, Fullerton Fire Dept. logo. Promo.

3107 BERLIET AIRPORT FIRE TRUCK 140 mm #351>#3351>1986-1990

Same model as #351, with red-hub wheels, white Sapeurs Pompiers logo.
1. Red body.

3108 MERCEDES SNORKEL FIRE TRUCK 152/175 mm 1986-1987

Similar model to #3105, with cast body, matching mount, black chassis, blue plastic windows nad dome lights, white arms, red basket, wheels with red hubs, Service Departemental decals.
1. Red body.

3109 BERLIET LADDER TRUCK 141 mm #352>#3352>1987

Same model as #352, presumably with red-hub wheels.
1. Red body.

3110 GMC FIRE WRECKER 156 mm 1/50 1987-1989

Open truck with cast cab, rear body and chassis, tan plastic cab roof, black seats, grille and boom, wheels with red hubs, spare wheel, white Secours Routiers No. 3 logo.
1. Red cab, body and chassis.

3111 MERCEDES-BENZ LADDER TRUCK 150 mm 1/50 #361>#3361>#3101 >1988-

Same truck as #3108, with red mount, unpainted aerial ladder, wheels with red hubs, Service Departemental decals.
1. Red body.

3112 BERLIET SNORKEL FIRE TRUCK 142/175 mm 1988-

Same truck as #3109, with same mount, arms and basket as #3108, wheels with red hubs, white Sapeurs Pompiers logo.
1. Red body.

3113 GMC FIRE DEPT. TRUCK 131 mm 1/50 1988

Open truck with cast cab, rear body and chassis, bue plastic windows and dome light, black grille, wheels with red hubs, spare wheel, Service Departemental de Secours de Loir & Cher logo.
1. Red cab, body and chassis.

3114 MERCEDES-BENZ FIRE ENGINE 125 mm 1988-

Truck with cast cab and rear body, blue plastic windows and dome lights, silver gray roof and panels, silver ladders, black hose, nozzle, front fenders and inner chassis, wheels with red hubs, white stripes, no logo.
1. Red cab and body.

3115 GMC SMOKE EJECTOR 128 mm 1/50 1989-1990

Same cab and chassis as #3113, with red plastic rear body, silver gray smoke ejector, same logo as #3113.
1. Red cab and chassis.

3116 GMC SMOKE EJECTOR 126 mm 1/50 1990-

Same cab, chassis and silver gray roof as #3110, with same rear body and ejector as #3115, white St. Tropez logo.
1. Red body and chassis.

3117 GMC FIRE WRECKER 156 mm 1/50 1990-
Same cab and chassis as #3115, with same rear body and boom as #3110, white S.D.I. du Var logo.
1. Red cab, body and chassis.

3118 IVECO FIRE ENGINE 128 mm 1990-
Truck with cast cab, rear body and chassis, blue plastic windows and dome light, silver ladders, silver gray panels, rear floor and inner chassis, black front fenders and grille, red and white hose reels, wheels with red hubs, wheel, white phone and 18 logo.
1. Red cab, body and chassis.

3119 SIDES AIRPORT FIRE TRUCK 181 mm 1/63 3/1991-
Truck with cast body, tinted plastic windows and dome lights silver gray panels, white nozzle and floor, black chassis, fast wheels, white stripes, S2000 abd Aeroports De Paris logo.
1. Red body.

3120 SIDES AIRPORT FIRE TRUCK 181 mm 1/63 3/1991-
Same basic model as #3119, but with black panels and Aeroport de Berlin-Tegel logo.
1. Red body.

3121 GMC COVERED FIRE TRUCK 131 mm 1/50 1/1991
Same truck as #3113 plus silver gray plastic cover, white grille, Sapeurs Pompiers and coat of arms decals.
1. Red cab. body and chassis.

3122 MERCEDES FIRE VAN & BOAT TRAILER 192 mm 1/50 1992-
Same model as #2132, with white Sapeurs Pompiers logo, pulling red cast trailer with red-hub wheels and red plastic boat.
1. Red body.

3300 SERIES

This series was created in 1980, chiefly by adding a 3 to the numbers of numerous 300 series trucks. Few new models were added, all in 1980 and all with low numbers, and in 1985 the last of the 3300 series were leaving the scene. Most of them, in fact, had been dropped at the end of 1981 or 1982, some reappearing later with new numbers. So there is not much news to report here, except that most models had more-or-less fast wheels by now.

3300 SAVIEM WRECKER 105 mm 1/50 #366>1980-1985>#2006
Same model as #366.
1. White cab, red chassis and rear body.

3301 SAVIEM FIRE WRECKER 105 mm 1/50 #366B>1980-1981>#2102
Same model as #366B.
1. Red body.

3302 SAVIEM POLICE WRECKER 105 mm 1/50 #366C>1980
Same model as #366C.
1. Blue body.

3303 SAVIEM FIRE ENGINE 114 mm 1/50 7/1980-1982>#2001>#2101
Truck with cast cab, rear body, blue plastic windows and dome lights, silver ladder, white hose, red and white hose reel, black chassis, red hubs, tires, Sapeurs Pompiers and coat of arms.
1. Red cab and body.

3304 FIRE JEEP & PUMP TRAILER 131 mm 1/43 4/1980-1982>#2114
Jeep with cast body and matching folding windshield frame, plastic windshield, gray top and chassis, pulling cast #360 trailer, both with red wheels, tires, white Service Departemental d'Incendie decal on hood.
1. Red bodies.

3305 BERLIET LOW LOADER 336 mm 1/67 #305B>1980-1981
Same model as #305B, including red Renault van, GDF logo.
1. Red cab and semi, silver cab chassis.

3306 SKIP DUMPER 69 mm 1/43 5/1980-1982
Dumper with cast body, matching plastic tipper and chassis, black plastic seat and steering wheel, yellow hubs, tires.
1. Yellow body, chassis and tipper.

3307 IVECO COVERED TRUCK 113 mm 5/1980-1982>#3007
Same model as #3007, with Esso Aseptogyl decals.
1. White cab and body, silver chassis, white cover.

3321 SAVIEM CAR CARRIER & TRAILER mm #321>1980-1981
Same model as #321. Used later in #7006.
1. Blue cab, upper decks and ramp, silver lower decks.

3350 BERLIET FIRE ENGINE 123 mm #350>1980-1981>#2107
Same model as #350.
1. Red body.

3351 BERLIET AIRPORT FIRE TRUCK 140 mm #351>1980-1981>#3107
Same model as #351.
1. Red body.

3352 BERLIET LADDER TRUCK 141/163 mm #352>1980-1981>#3109
Same model as #352.
1. Red body.

1. Red body.

3353 RICHIER CRANE TRUCK 139 mm #353>1980-1981>#3102
Same model as #353.
1. Yellow body and boom.

3354 BERLIET FOREST FIRE TRUCK & TRAILER 145 mm #354>1980-1981>#2103
Same model as #354.
1. Red truck and trailer bodies.

3355 PEUGEOT J7 MINIBUS 92 mm 1/50 #355>1980
Same model as #355.
1. Blue body, white roof.

3356 VOLVO DUMP TRUCK 160 mm #356>1980-1981>#3103
Same model as #356.
1. Yellow body.

3357 UNIC SAHARA DUMP TRUCK 158 mm #357>1980
Same model as #357.
1. Yellow body.

3358 MERCEDES OVERHEAD SERVICE TRUCK 149/175 mm #358>1980-1982>#3105
Same model as #358.
1. Orange body, yellow arms and basket.

3359 SIMCA-UNIC SNOWPLOW TRUCK 142 mm #359>1980>#2108
Same model as #359.
1. Yellow body, orange cover.

3361 MERCEDES LADDER TRUCK 150/163 mm #361>1980-1981> #3101>#3111
Same model as #361.
1. Red body.

3362 HOTCHKISS FIRE ENGINE 111 mm #362>1980-1981>#2000> #2100
Same model as #361.
1. Red body.

3363 MAGIRUS COVERED SEMI 250 mm #363>1980-1981
Same model as #363.
1. Metallic red cab, silver chassis, yellow semi.

3364 MERCEDES BUCKET TRUCK 136 mm #364>1980-1981
Metallic red cab, silver rear body, yellow bucket.

3365 INTERNATIONAL POWER SHOVEL #365>1980
Same model as #365.
1. Yellow and white body parts.

3367 VOLVO SHOVEL LOADER 129 mm #367>1980-1982>#3104
Same model as #367.
1. Yellow body and shovel.

3368 CITROEN C35 AMBULANCE 96 mm #368>1980-1981>#2002>#2116
Same model as #368.
1. Red body, white roof.

3369 DAF TANKER SEMI 217 mm #369>1980-1985>#3500
Same model as #369, with Shell logo.
1. Yellow cab and semi, silver cab chassis, white tank.

3370 SAVIEM SEMI 218 mm #370>1980-1982>#3502
Same model as #370, with Renault logo.
   1. Yellow cab and semi bed, white box.

3371 CITROEN AMBULANCE & LIFEBOAT TRAILER 192 mm #371>1980-1981
Same model as #371.
   1. Red body, white roof, brown and gray boat.

3372 PEUGEOT J7 POLICE BUS 91 mm #372>1980-1982>#2003
Same model as #372.
   1. Blue body, white roof.

3373 MERCEDES LIVESTOCK TRUCK 120 mm #373>1980-1981
Same model as #373.
   1. Green cab and rear body, silver chassis.

3374 IVECO DUMP TRUCK 135 mm #374>1980-1981>#2009>#2109
Same model as #374.
   1. Yellow cab, red tipper, black chassis.

3375   BERLIET FIRE ENGINE   116 mm   #375>1980-1982>#2008>#2106
Same model as #375.
   1. Red body.

3376 MERCEDES BULK CARRIER SEMI 211 mm #376>1980-1981>3505
Same model as #376.
   1. Red cab, silver semi, white tank.

3378 MERCEDES EXCAVATOR TRUCK 173 mm #378>1980-1981
Same model as #378.
   1. Orange body and other parts.

3379 MERCEDES GARBAGE TRUCK 138 mm #379>1980-1982
Same model as #379.
   1. Brown cab, black chassis, tan rear body.

3380 PEUGEOT J7 FIRE AMBULANCE 93 mm #380>1980>#2126
Same model as #380.
   1. Red body, white roof.

3384 MERCEDES COVERED TRUCK 121 mm #384>1980-1981
Same model as #384.
   1. Silver cab, blue rear body, red chassis, light blue cover.
   2. Red cab, yellow rear body, Transport Ecallard logo. Promo.

3385 SAVIEM HORSE VAN 229 mm #385>1980-1982
Same model as #385.
   1. Silver cab and semi bed, brown semi body.

3386 MERCEDES PROPANE TANKER 127 mm #386>1980-1982
Same model as #386.
   1. Metallic blue cab and rear body, white tank.

3388 SAVIEM STAKESIDE SEMI 229 mm #388>1980-1982
Same model as #388.
   1. Green cab and semi bed, white stake body.

3391 DODGE FIRE TRUCK & TRAILER 180 mm #391>1980
Same model as #391. Truck later used as #2007, #2105, with pump trailer as #3004.
   1. Red truck and trailer bodies.

3500 FARM SERIES

In 1980 the agricultural models were renumbered 3510 to 3516 in the usual way, by prefixing a 3 to their previous numbers. No models were changed, and no new models were added.

3510 RENAULT FARM TRACTOR 109 mm #510>1980-1981

3511 TIPPING FARM TRAILER 180 mm #511>1980-1981

3512 RENAULT TRACTOR & TIPPING TRAILER 282 mm #512>1980-1981

3513 TANK TRAILER 140 mm #513>1980-1981

3514 RENAULT TRACTOR & TANK TRAILER 243 mm #514>1980-1981

3515 HARVESTER TRAILER 98 mm #515>1980-1981

3516 SPRAYER TRAILER 105 mm #516>1980-1981

3500 SEMI SERIES

In 1985, several years after the farm machinery had left the scene, new 3500 numbers were used for a series of semi-trailer trucks, the first of which were taken from the 3300 series or adapted from 3300 models. I assume this was done to put these models in a higher price bracket. And I must admit that #3507 is a truck and trailer, not a semi! As for the American term "semi" itself--well, it's short!

3500 DAF TANKER SEMI 220 mm #369>#3369>1985-1986
Same basic model as #269, with cast cab, chassis and semi, white plastic tank and airscreen, plastic cab floor, catwalk and ladder, silver horns and stacks, wheels with various hubs, logo decals or labels.
1. White cab and catwalk, yellow semi and cab floor, red hubs, red and yellow Shell decals.
2. White cab, floor, airscreen, tank and catwalk, silver chassis, semi and hubs, red-white-black Texaco labels.
3. Blue cab and semi, red chassis and catwalk, black cab floor, white airscreen with Solido label, black-red-blue Elf Antar labels.

3501 DAF COVERED SEMI 259 mm #363>#3363>1985-
Same basic model as #363, with cab components as #3500, cast semi with plastic cover and black chassis, wheels with hubs, logo decals or labels.
1. Yellow cab, airscreen and semi, black floor and chassis, blue cover, yellow Danzas decals.
2. Yellow cab and semi, silver chassis, black floor, cover and airscreen with Solido label, multicolored Kvas labels.
3. Yellow cab and semi, silver chassis, blue floor and hubs, white airscreen, gray cover, blue-yellow-white Opal labels.
4. Red cab and semi, white airscreen and cover, Ipone logo.
5. White cab, semi and cover, Texaco Lubricants logo. Promo.

6. White cab and semi, blue cover, Feller logo. Promo.
7. Blue cab and semi, yellow cover, Michelin logo. Promo.

3502 SAVIEM BOX SEMI 222 mm #370>#3370>1985-1986

Same basic model as #370, with cast cab, chassis, semi bed, plastic box, airscreen, semi chassis, silver horns and stacks, wheels with hubs, logo labels.
1. Red cab and bed, black chassis, box, and airscreen with Solido label, red-white-yellow-brown-black Renault labels.
2. Blue cab and bed, white chassis, airscreen and box, yellow hubs, multicolored C. Saunders Ltd. labels.
3. Red cab and chassis, white box, GB inno-bm logo. Promo.

3503 SAVIEM STAKE SEMI WITH PIPES 226 mm #388>#3388>1985

Same basic model as #388, but with black pipes instead of logs. Cast cab and semi bed, black chassis, white plastic airscreen with Solido label and stake body, wheels with hubs, logo labels.
1. Green cab and bed, yellow-orange-green-black Eurotube labels.

3504 IVECO FERRARI SEMI 228 mm 1985

Semi with cast cab, chassis, semi bed, unpainted inner chassis, plastic windows, silver horns and stack, red semi body, opening door and inner chassis, wheels with hubs, yellow and black Ferrari decals.
1. Red cab and semi body, black chassis.

3505 MERCEDES BULK CARRIER SEMI 209 mm #376>#3376>1985

Same basic model as #376, with cast cab, chassis, semi body, unpainted inner chassis, silver plastic horns and stack, white tank, red catwalk and ladder, wheels with hubs, black stripe labels, Mercedes star/German flag tampo-print on cab.

1. Red cab, black chassis, yellow semi.

3506 IVECO BULK CARRIER SEMI 208 mm 1985-1986

Same cab as #3504, same semi as #3505, with white catwalk and ladder, blue hubs, logo decals or labels.
1. Blue cab and semi, white chassis and tank, blue and white Milk labels.
2. Orange cab, chassis and tank, blue semi and hubs, blue and white Tecni Plast decals.

3507 DAF COVERED TRUCK AND TRAILER 249 mm 1985-

Covered truck and trailer with cast cab, truck and trailer bodies, black chassis and hitch, plastic windows, airscreen and covers, silver horns and stacks, wheels with hubs, logo labels.
1. Red cab and bodies, white airscreen and covers, red-white-black Ipone labels.
2. Red cab and bodies, yellow airscreen and covers, Transports Solido logo.
3. Yellow and blue, La Poste logo.

3508 MACK SEMI-TRAILER VAN 247 mm 1985-

Van semi with cast cab, chassis, semi bed, plastic windows, semi body, matching opening doors and inner semi chassis, silver horns and stack, black inner cab chassis, wheels with hubs, logo labels
1. White cab and semi body, blue cab chassis and semi bed, yellow and blue Team Husqvarna labels plus emblem tampo on hood.
2. Metallic red cab, green semi bed, silver chassis, dark blue semi body, red-yellow-white-black Maffucci Mayflower labels.
3. White cab and semi body, blue chassis and bed, red-white-blue Sealink logo.
4. Black cab and semi, Texaco logo. Promo.
5. White and orange, Multiplus logo. Promo.

3509   RENAULT FIRE TANK SEMI   216 mm   1986-

Tanker with cast cab, black chassis and semi, unpainted inner chassis, plastic windows, tank and matching ladder, white airscreen, catwalk and inner semi chassis, silver horns and stacks, wheels with red hubs, white Sapeurs Pompiers decals
1. Red cab and tank.
2. Orange cab and semi, Seca logo. Promo.
3. Black and tan, Q8 Koweit Petroleum logo. Promo.

3510   MERCEDES CONTAINER SEMI   228 mm   1986-

Semi with cast cab, semi bed, black chassis, plastic windows, container with matching opening doors, black cab floor and parts, wheels with hubs, logo labels.
1. Red cab and container, red hubs, red-white-black Bridgestone logo.
2. Green cab and container, green hubs, multicolored Andros logo.
3. Green cab, semi bed and hubs, white container, green-black-white Les Déménageurs Bretons logo.
4. Orange cab, metallic gray chassis and semi, Kunz logo. Promo.
5. Red, black and white, Brico logo in 4 languages. Promo.
6. White, A. Germain logo. Promo.
7. White and blue or white and black, France Routes logo. Promo.
8. Red and black, France Routes logo. Promo.
9. Yellow and black, Michelin Competition logo. Promo.
10. Blue, France Telecom logo. Promo.

3511   MACK FIRE SEMI   250 mm   1987-

Semi with cast cab, silver chassis, plastic windows, semi body and black bed, black inner cab chassis, silver horns and stack, wheels with red hubs, gold and black F.D.N.Y. decals and control panel labels.
1. Red cab and semi body.

3600 SERIES

This small series of snow-oriented vehicles was introduced in 1986 and includes only seven models. In 1980, though, the #650 farm set was renumbered #3650, and we might as well list it here.

3600   MERCEDES UNIMOG SNOWPLOW & TRAILER   204 mm   #3005>1986

Same model as #3005.
1. Yellow bodies, orange cover.

3601   KÄSSBOHRER PISTENBULLY   89/163 mm   1/50   9/1986-

Snowmobile with cast cab and chassis, plastic windows and dome light, black interior, grille, stack, hubs and plows, gray tracks, K monogram decals. Usually with front and rear plows, but also exists without them.
1. Dark orange cab, silver chassis, with plows.
2. Light orange cab, silver chassis, without plows.

3602   KÄSSBOHRER PISTENBULLY EXPEDITION   132 mm   1987-

Snowmobile with same cab, chassis, parts and front plow as #3601, plus white plastic body with windows, K and stripe labels.
1. White cab and body, black chassis.

3603   KÄSSBOHRER PISTENBULLY GRITSPREADER   120 mm   1987-1989

Snowmobile with same cab, chassis and parts as #3601, no plow, plus red plastic gritspreader, K decals on cab.
1. Red cab, black chassis.

3604   UNIC SNOWPLOW TRUCK WITH GRITSPREADER   mm   1987

Same truck as in #3600. No other data.

3606 MERCEDES UNIMOG WITH SNOWPLOW 134 mm 1988-

Truck with cast body, blue plastic windows and dome light, black chassis, plow, mount and stack, wheels with chains, spare wheel, red and white stripes tampo-printed on hood.

    1. Orange-red body.

3607 FIRE RESCUE SNOWMOBILE 92 mm 1990-

Snowmobile with same cab, chassis and parts as #3601, no plow, plus red plastic body with windows, exhaust stack, dome light, Rettungsdienst and Notarzt labels.

    1. Red cab, black chassis.

3650 FARM SET #650>1980-1981

Same set as #650. Models in standard colors.

3700 SERIES

Series? The only 3700 model was the Renault car carrier, which has been seen under other numbers before and since.

3700 RENAULT CAR CARRIER & TRAILER 425 mm #321>#3321>1984-1985>#7006

Same model as #321. Colors may be as before.

3800 SERIES

The few helicopters from the old 300 series were moved into a series of their own in 1980, and some of their successors are still in production. There have, in fact, been twenty-two of them to date, though only five different makes are represented.

3810 GAZELLE HELICOPTER 171 mm 1/55 #381A>1980-1982
Same model as #381A, with orange and blue Europ Assistance and orange and black tailfin decals.
1. White body.

3811 GAZELLE HELICOPTER 171 mm 1/55 #381B>1980-1982
Same model as #381B, with white tailfin and Gendarmerie decals.
1. Blue body.

3812 GAZELLE HELICOPTER 171 mm 1/55 #381A>1980-1981
Same model as #381A, with Italian roundels and Carabinieri   decals.
1. Blue-black body.

3813 GAZELLE HELICOPTER 171 mm 1/55 #381B>1980-1981
Same model as #381B, with Italian roundels and Polizia decals.
1. Dull blue body.

3814 ALOUETTE SECURITY HELICOPTER 178 mm 6/1980-
Helicopter with cast body, blue plastic windows, silver interior, black motor, main rotor and wheels, white tail, rotor and struts.
1. Red body, Sécurité Civile decals.
2. Red body, Air Zermatt logo. Promo.
3. Red body, Rega logo. Promo.
4. Yellow body, OAMTC Christophorus logo. Promo.

3815 GAZELLE ARMY HELICOPTER 171 mm 1/55 10/1980-
Same basic model as #3810, plus olive launchers and white missiles, no decals.
1. Olive body.
2. White body with olive camouflage.

3816 ALOUETTE GENDARMERIE HELICOPTER 178 mm 1985-1986
Same basic model as #3814, presumably blue.

3817 ALOUETTE ASI HELICOPTER 178 mm 1986-1987
Same basic model as #3810, with yellow tail and struts, red Air Secours International logo.
1. White body.

3818 ALOUETTE TELE-UNION HELICOPTER 178 mm 1986
Special issue not readily available. No data.

3819 GAZELLE GENDARMERIE HELICOPTER 171 mm 1/55 1986-1988
Similar to #3811, with white Gendarmerie and tail decals; also used in #7028 set.

3820 ALOUETTE SECURITY HELICOPTER 178 mm 1986-1989
Similar to #3814; used in #7015 set.
1. Red body.

3821 ALOUETTE PRIVATE HELICOPTER 178 mm 1987-1988
Same basic model as #3814, with amber windows, yellow-orange tail and struts, red-orange-yellow trim decals.
1. Black body.

3822 GAZELLE CIVIL HELICOPTER 171 mm 1/55 1988-1991
Same basic model as #3810, with black decals.
1. Yellow body.

3823 ALOUETTE GENDARMERIE HELICOPTER 178 mm 1988-1991
Successor to #3816, with blue windows, white tail and struts, French roundel and white JBL Gendarmerie decals.
1. Blue body.

3824 PUMA ARMY HELICOPTER 181 mm 1988-
Helicopter with cast body, plastic windows, black interior, rotors and wheels, French roundel and white lettering decals.
1. Olive body.
2. Olive body, Aerospatiale logo.

3825 PUMA CIVIL HELICOPTER 181 mm 1988-
Same basic model as #3824, with white interior and struts, two decal types.
1. Red body, black Aerospatiale decals.
2. Red body, white Super Puma as 332 decals.
3. White body, Rajavartiolaitos logo. Promo.

3826 ALOUETTE ADP HELICOPTER 178 mm 1991-
Same basic model as #3823, with white tail and struts, green and black logo.
1. White body.

3827 GAZELLE GENDARMERIE HELICOPTER 171 mm 1/55 1991-
Successor to #3819, with French roundel and white lettering and tailfin.
1. Blue body.

3828 COUGAR ARMY HELICOPTER 181 mm 1992-
Same basic model as #3824 Puma, with French roundels, white rockets and lettering.
1. Olive body.

3829 GAZELLE HOT HELICOPTER 171 mm 1/55 1992-
Same basic model as #3810, with black parts, French roundel and white lettering.
1. Tan body and rockets.

3830 AGUSTA HELICOPTER ___ mm 1992-
Helicopter with cast body, plastic windows, black rotors and wheels, white lettering.
1. Red body.

3831 AGUSTA HELICOPTER ___ mm 1992-
Same basic model as #3830. May be army version.
1. Green body.

4000 SERIES

By 1980 it was obvious that the classic oldtime cars in the 10 and 100 series belonged in a class by themselves numerically, and the 10 series models now had 40 prefixed to their catalog numbers. Since then, numerous models have been added to this series, though there are still many gaps in the numbering.

4000 CITROEN 15CV PARIS-MOSCOW 110 mm 1/43 1985
Special version of #4032. No data. Is number genuine?

4001 MERCEDES-BENZ SSKL 1931 100 mm 1/43 4/1980-1982
Open sports car with cast body, chassis, plastic windshield, fenders, black interior, silver gray drive train and exhaust pipes, silver grille and lights silver or red spoked hubs, two spares, hood strap.
　　1. Silver body and chassis, silver gray fenders.
　　2. White body, chassis and fenders.

4002 JAGUAR SS 100 1938 92 mm 1/43 4/1981-
Convertible with cast body, fenders, plastic windshield, black interior and folded top, silver grille, lights, chassis and spoked wheel hubs, spare wheel, hood strap.
　　1. Cream body, red fenders.
　　2. Red body and fenders.
　　3. Green body and fenders.
　　4. Green body, black fenders.

4003 TALBOT T23 1937 110 mm 1/43 6/1981
Convertible with cast body, matching opening doors, chassis, plastic windshield, interior and matching folded top, silver grille, lights, bumpers, drive train and spoked wheel hubs.
　　1. Pale green body, dark green chassis, black interior.

　　2. Pale green body, black chassis and interior.
　　3. Pale blue body and chassis, brown or black interior.
　　4. Bronze body, black chassis and interior.
　　5. Metallic charcoal gray body and chassis, light brown interior.

4004 MERCEDES-BENZ SSKL MILLE MIGLIA 100 mm 1/43 12/1981-
Same basic model as #4001, with red fenders, silver drive train and exhaust pipes, black board with silver cases on right, black #87 decals.
　　1. Silver body and chassis.
　　2. White body and chassis.

4005 ALFA ROMEO 1931 (not issued)

4031 DELAGE D-8120 1938 122 mm 1/43 #31>1980-1987
Same model as #31, with tan interior, silver parts. Standard colors plus:
　　1. White body, black chassis and raised top.
　　2. Red body, black chassis and raised top.
　　3. Red body, black chassis and top, black lion and SI EZY/EURE logo. Promo.
　　3. Dark gray body, black chassis and raised top.

4032 CITROEN 15CV 1939 110 mm 1/43 #32>1980-
Same model as #32 (with cast-in rear spare).
　　1. Cream body, hoods and hubs.
　　2. Light gray body and hoods, cream hubs.
　　3. Black body and hoods, cream hubs.
　　4. Silver body and hoods, red hubs, 1938-1988 decals.
　　5. Cream body, hoods and hubs, 1938-1988 decals.
　　6. Gray-blue body, B E P logo. Promo.
　　7. Black body, Bendix-France logo. Promo.

4033 CITROEN 15CV FIRE CHIEF 110 mm 1/43 #32A>1980-1986
Same model as #32A, with Sapeurs Pompiers decals.
  1. Red body.

4034 CITROEN 15CV FFI 110 mm 1/43 #32B>1980-1982
Same model as #32B, with FFI decals.
  1. Olive body with white green camouflage.

4035 DUESENBERG SPIDER 1931 130 mm 1/43 #35>1980-1987
Same model as #35.
  1. Light blue body, black chassis and top, brown interior.
  2. Dark blue body, black chassis and top, brown interior.
  3. Purple body, black chassis and top, brown interior.

4036 BUGATTI ROYALE 1930 135 mm 1/43 #136>#4136>1983-
Same model as #136, but with blue interior and apparently non-removable hood.
  1. Black body.
  2. Black body, Auto Festival 1987 logo. Promo.
  3. Black body, Rallye Européen logo. Promo.

4037 PACKARD SUPER EIGHT 1937 138 mm 1/43 1983-1988
Closed convertible with cast body, matching opening doors, chassis, plastic windshield, interior, black top and trunk, silver grille, lights and bumpers, wheels with silver hubs, twin side spares.
  1. White body, black chassis, gray interior.
  2. Yellow body, black chassis, brown interior.
  3. Light green body, black chassis, gray interior.
  4. Olive green body, black chassis, blue interior.

4038 CADILLAC FIRE BRIGADE CAR 130 mm 1/43 1984-1985
Sedan with cast body, chassis, plastic windows, white ambulance interior, black roof panel, silver grille, lights and bumpers, wheels with spoked hubs, twin side spares, yellow Manhattan Fire Brigade (sic) and Statue of Liberty tampo-print.
  1. Red body and chassis.

4039 CITROEN 15CV FFI 110 mm 1/43 #32B>#4034>1984-1985
Apparently a reissue of #4034 in military form.

4040 CITROEN 15CV FFI 110 mm 1/43 1984-1985
Black version of #4034 with wider white FFI lettering.
  1. Black body.

4041 CITROEN 15CV TAXI 110 mm 1/43 1984-1985
Same basic model as #4032, plus yellow sign with black Taxi lettering.
  1. Blue-black body.

4042 CADILLAC AMBULANCE 130 mm 1/43 1986-1987
Same basic model as #4038, with tan ambulance interior, orange Denver + First Aid & Rescue Squad tampo-print.
  1. White body and chassis.

4043 CADILLAC POLICE CAR 130 mm 1/43 1987-1988>#4057
Same basic model as #4038, with tan ambulance interior, black and white stripes, badge and 101 Police decals.
  1. Black body and chassis.

4046 ROLLS-ROYCE PHANTOM III 1939 129 mm 1/43 #46>1980-1988
Same model as #46.
  1. Metallic gray body, black chassis, white top, gray interior.
  2. Silver body, black chassis, Moerkerke logo. Promo.

4047   PACKARD SEDAN 1937   131 mm   1/43   1986-
Sedan with cast body, matching opening front doors, chassis, plastic windows, light tan or gray interior, silver grille, lights and bumpers, wheels with silver hubs, twin side spares.
1. Cream body, metallic brown chassis.
2. Dark red body, black chassis.
3. Light orange body, black chassis.
4. Blue body, black chassis.
5. Black body and chassis.

4048   DELAHAYE CONVERTIBLE 1937   117 mm   1/43 #48>1148>1980
Same model as #48. Standard colors plus:
1. Red body, white chassis, gray top and interior.
2. Yellow body, black chassis, white top and interior.
3. Dull blue body and chassis, white top and interior.

4051   DELAGE COUPE DE VILLE 1938   122 mm   1/43   #51>#1151> 1980-1986
Same model as #51. Standard colors plus:
1. Cream yellow body, brown chassis, tan interior.
2. Light tan body, pinkish-brown chassis, tan interior.

4053   DELAGE CONVERTIBLE 1938   (top down, not issued)

4055   CORD L29 CONVERTIBLE SEDAN 1930   116 mm   1/43 #55>1980-
Same model as #55. Standard colors plus:
1. Silver body, red chassis, black roof, gray interior.
2. Pale blue body, black chassis and roof, gray interior.
3. Blue body and chassis, black roof and interior.

4057   CADILLAC POLICE CAR   130 mm   1/43   1988
Same basic model as #4043, plus white roof panel, silver siren on hood, white star with Police No. 3 County.
1. Black body.
2. Black body, white chassis?
3. Pink body, Lorenzi Modellismo logo. Promo.

4059   RENAULT 40 CV 1925   123 mm   1/43   #59>#1159>1980-
Same model as #59. Standard colors plus:
1. White body and chassis, black roof and interior, brown hubs.
2. White body, black chassis, roof and interior, silver hubs.
3. Orange body, black chassis, top and interior.
4. Yellow body, black chassis, top and interior.
5. Red-brown body, black chassis, top and interior.

4060   CADILLAC CADBURY'S VAN   130 mm   1/43   1985
Van with cast body, chassis, plastic windows, black roof panel, tan interior, silver grille, lights, bumpers and wheel hubs, twin side spares, multicolored Cadbury's Cocoa Essence logo.
1. Light yellow body, blue chassis.

4061   CADILLAC BANANIA VAN   130 mm   1/43   1986
Same model as #4060, but with multicolored Banania logo.
1. Yellow body, metallic brown chassis.

4062   HISPANO-SUIZA PHAETON 1926   114 mm   1/43 #62>#1162>1980-1985
Same model as #62. Standard colors.

4065   CADILLAC WATERMAN VAN   130 mm   1/43   1988-1989
Same model as #4060, but with multicolored Waterman Ink logo.
1. Yellow body, black chassis.

4067 MERCEDES-BENZ 540K 122 mm 1/43 #67>1980-1988
Same model as #67. Standard colors plus:
  1. Red body and chassis, white top, black interior.
  2. Red body and chassis, black top and interior.

4070 CADILLAC FIRE VAN 130 mm 1/43 1990-
Same basic model as #4060, but with silver siren on hood, star with Fire Chief lettering decals on doors. Successor to #4075.
  1. Red body, black chassis.

4071 ROLLS-ROYCE PHANTOM III TOWN CAR 130 mm 1/43 #71>1980-
Same model as #71. Standard colors plus:
  1. Red body, black chassis, gray interior.
  2. Dark green body and chassis, tan interior.
  3. Black body and chassis, gray or tan interior.
  4. Black body and chassis, F R on doors. Promo.
  5. Black body and chassis, Rallye Européen logo. Promo.

4075 CADILLAC FIRE VAN 130 mm 1/43 1987-1988
Same model as #4060, but with black and gold Sellers Fire Dept. and No. 1 F.D. badge decals.
  1. Red body, black chassis.

4077 ROLLS-ROYCE PHANTOM III DROPHEAD 130 mm 1/43 #77>1980-
Same model as #77. Standard colors plus:
  1. White body, black chassis and folded top, gray interior.
  2. Cream body, black chassis and folded top, tan interior.
  3. Greenish-blue body, black chassis and folded top, gray   interior.
  4. Metallic olive green body, black chassis and folded top, gray interior.
  5. Metallic silver blue body, black chassis and folded top, gray interior.

  6. Metallic blue body, black chassis and folded top, gray interior.
  7. Metallic gray body, black chassis, tan interior.

4078 DELAHAYE CABRIOLET 1939 118 mm 1/43 #78>1980-1985
Same model as #78. Standard colors plus:
  1. Blue body, white chassis, black interior and folded top.

4080 CORD l29 COUPE 1931 117 mm 1/43 #80>1980-
Same model as #80. Standard colors plus:
  1. Yellow body, black chassis, white roof.
  2. Orange body, black chassis, cream roof.
  3. Metallic green body, black chassis, white roof.
  4. Red-brown body, black chassis, white roof.
  5. Brown body, black chassis and roof.

4085 CADILLAC IMPERIAL LANDAULET 1931 129 mm 1/43 #85>1980-
Same model as #85. Standard colors plus:
  1. Red body, black chassis, roof and folded top.
  2. Light yellow body, black chassis, roof and folded top.
  3. Metallic gray-green body and chassis, white roof and folded top.
  4. Metallic bronze body and chassis, white roof and folded top.

4086 MERCEDES-BENZ 540K CABRIOLET 1939 123 mm 1/43 1988-
Open convertible with cast body, chassis, plastic windshield, black interior and folded top, silver grille, lights and bumpers, wheels with wire hubs.
  1. Red body, black chassis.
  2. Blue body, dark blue-gray chassis.
  3. Blue body, black chassis.
  4. Blue body, black chassis, Rallye Européen logo. Promo.

4088 BUGATTI ATALANTE 1939 109 mm 1/43 #88>1980-
Same model as #88. Standard colors plus:
1. Red body and chassis, black trim.
2. Dark red body and chassis, black trim.
3. Black body and chassis, light yellow trim.
4. Black body and chassis, blue trim.

4097 RENAULT REINASTELLA 1934 122 mm 1/43 #97>1980-
Same model as #97. Standard colors plus:
1. Gold body, black chassis.
2. Dark blue body, black chassis.
3. Navy blue body and chassis, tan roof panel, cream side panels to simulate wickerwork.
4. Black body and chassis, tan roof panel, cream side panels to simulate wickerwork.

4099 PACKARD CONVERTIBLE 1937 138 mm 1/43 1988-
Open convertible with cast body, matching opening doors, chassis, plastic windshield, gray interior, black trunk and folded top, silver grille, lights and bumpers, wheels with silver hubs and whitewalls.
1. Red body, black chassis.
2. Metallic dark gold body, metallic brown chassis.

#? CADILLAC PROMOTIONAL VANS 130 mm 1/43 year?
The following promotional versions of the Cadillac Van are known to exist. All have tan interior, silver grille, lights and bumpers, and wheels with hubs.
1. White body, red chassis, FEG Pieces Pour Freins logo.
2. White body, dark blue chassis, Expo '92 logo; 1992.
3. Yellow body and hubs, black roof panel and chassis, black and red XXXX The Popular Beer Castlemaine Perkins logo.
4. Yellow body, red chassis, Kodak Film logo. Promo.
5. Yellow body, red chassis, Nikon logo. Promo.
6. Yellow body, pale green chassis, black roof panel, green and yellow Queru logo.
7. Yellow body, pale green chassis, Goudkuipje logo.
8. Yellow body, black chassis, Fan Auto logo.
9. Maroon body and chassis, gold Hewlett-Packard Analytical Products logo.
10. Red body, gold chassis, Cadillac Eldorado Club logo. Promo.
11. Blue body, orange chassis, Groupe Azur logo. Promo.
12. Black body and chassis, white Zeller's logo.
13. Black body, Champagne Dravigny logo. Promo.

4100 SERIES

The 4100 series began as a home for the 100 series Age d'Or models when the new numbering system was introduced in 1980. Beginning in 1987, a few similar models have been added to the series, but only a few 4100 models remain in production today.

4102 CITROEN 15CV 1952 111 mm 1/43 1987>#4519
Later version of #32 with trunk instead of rear spare, otherwise essentially the same model.
1. Dark gray body.
2. Black body.
3. White body, Automobiles Anciennes de Troyes logo. Promo.
4. Ten different colors, Look and Like logo. Promo.

4109 BUGATTI ATALANTE OPEN TOP 1939 106 mm 1/43 1987-
Open convertible coupe with cast body, chassis, plastic windows, interior, black folded top and trim, silver grille and bumpers, wheels with wire hubs, whitewalls, rear spare.
1. Yellow body and chassis.
2. Blue body and chassis.

4115 CITROEN 15CV WITH GAS BOTTLES 111 mm 1/43 1987-1991
Same basic model as #1032, plus gray plastic roof rack with gas canisters.
1. Black body.
2. Ivory body, Cinquantenaire 15 CV logo. Promo.
3. Metallic gray body, Cinquantenaire 15 CV logo. Promo.
4. Black body, Automobile Club Lorrain Nancy logo. Promo.
5. Black body, Rallye Européen logo. Promo.

4132 MERCEDES SS 1928 111 mm 1/43 #132>1980-1981
Same model as #132. Standard colors.

4136 BUGATTI ROYALE 1930 135 mm 1/43 #135>1980-1981
Same model as #135. Standard colors.

4137 MERCEDES SS TORPEDO 1928 111 mm 1/43 #135>1980-1982
Same model as #137. Standard colors.

4140 PANHARD 35CV LANDAULET 1925 120 mm 1/43 #140>1980-1981
Same model as #140. Standard colors.

4144 VOISIN 17CV 1934 106 mm 1/43 #144>1980-1982
Same model as #144. Standard colors.

4145 HISPANO-SUIZA 1926 115 mm 1/43 #145>1980-1985
Same model as #145. Standard colors.

4149 RENAULT 40CV 1925 123 mm 1/43 #149>1980-1986
Same model as #149. Standard colors plus:
1. Metallic gray body, blue-black chassis, black roof and top.

4154 FIAT 525N 1929 106 mm 1/43 #154>1980-1985
Same model as #154. Standard colors.

4156 DUESENBERG J 1931 128 mm 1/43 #156>1980-
Same model as #156. Standard colors plus:
1. Green body, black chassis and roof.
2. Metallic silver blue body, black chassis and roof.
3. Metallic dark blue body, black chassis and roof.
4. Metallic light purple body, black chassis and roof.
5. Metallic brown body, black chassis and roof.

4157 FIAT 521 CLOSED TOURER 109 mm 1/43 1990-1991
Same basic model as #154, but with raised black top.
   1. Yellow body, black chassis, top and interior.

4159 FORD V8 SEDAN 1936 107 mm 1/43 1992-
Sedan with cast body, chassis, plastic windows, black roof panel and grille, brown interior, silver radiator shell and bumpers, wheels with silver hubs, whitewalls.
   1. Brown body, black chassis.

4300 SERIES

Solido planned a series of cars of the fifties, but it never materialized. The following models were planned:

4300 Renault 4CV Sedan

4301 Renault 4CV Convertible

4302 Cadillac 1953 Convertible

4303 Hotchkiss 686

4400 SERIES

The two prewar bus models that would have begun a 400 series appeared in June and September of 1980 to begin the 4400 series, which has continued to this day and includes numerous models of prewar commercial vehicles.

4401 RENAULT 1936 PARIS BUS 186 mm 1/50 6/1980-1990
Singledeck bus with cast lower body, plastic windows, upper body, brown seats, black chassis, grille and parts, green hubs, tires, Banania decals.
1. Dark green lower, cream upper body.

4402 AEC 1939 DOUBLEDECK BUS 159 mm 1/50 9/1980-1990
Doubledecker with cast upper and lower body, plastic windows, white divider and pillar, brown interior and hubs, silver mirrors and radiator shell, black grille and chassis, tires,
1. Red body, red and black Hovis decals on yellow background on left, pink and brown Swan Vestas logo on right.
2. Red body, red-ywllow-blue-white Maples Furniture logo on left, red-white-blue British Airways logo on right.

4403 CITROEN C4F 1930 FIRE TRUCK 110 mm 1/43 12/1981-1987
Open-cab fire truck with cast hood-chassis, plastic windshield, red rear body and racks, red and white hose reels, brown ladders and seats, gold radiator shell and lights, black grille, silver running boards, wheels with red hubs, white Ville d'Argeles-Gazost decals.
1. Red chassis-hood and body.
2. Red chassis-hood and body, Springfield F.D. and Vancenase AQ logo. Promo.

4404 AEC 1939 DOUBLEDECK BUS 159 mm 1/50 9/1980-1990
Same basic model as #4402, with yellow divider and pillar, green chassis and interior, Brymay Matches decals of left, Littlewoods on right.
1. Green upper and lower body.

4405 CITROEN C4F 1930 HOTEL BUS 106 mm 1/43 12/1981-1985
Truck/bus with cast front and rear body, chassis, plastic windows, brown upper body and opening rear door, silver radiator shell, lights and running boards, black grille, wheels with hubs, Palace Hotel logo.
1. Yellow body, brown chassis and hubs, black-on-white logo.
2. Metallic green body, black chassis, silver hubs, gold logo plus monogram on doors.

4406 RENAULT 1936 LYON BUS 186 mm 1/50 2/1982-1985
Same basic model as #4401, with gray interior, brown hubs, light cream roof boards with Affichage Giraudy labels, destination labels and monogram decals below.
1. Red lower, light cream upper body.

4407 CITROEN C4 MICHELIN VAN 106 mm 1/43 6/1982-1985
Box van with cast front and rear body, chassis, plastic windows, brown rear box and opening doors, black grille, silver radiator shell, lights and running boards, wheels with yellow hubs, Michelin decals on box and door.
1. Yellow body, blue chassis.

4408 CITROEN C4 COAL TRUCK 106 mm 1/43 1983-1985
Same basic model as #4407, but with out rear box, with black rear sides-load of coal, hubs matching body, Alazard decals.
1. Red body, black chassis.
2. Tan body, brown chassis.

4409 CITROEN C4 SAMARITAINE VAN 106 mm 1/43 1983-1985
Same model as #4407, with blue box and hubs, Grand Magazins de la Samaritaine and letter S decals.
1. Blue body, dark blue chassis.

4410 CITROEN C4 WRECKER 110 mm 1/43 1984-1985
Wrecker with cast body, chassis, unpainted boom and mount, plastic windows, black hook and grille, silver radiator shell and lights, wheels with hubs matching body, spare wheel, logo decals. 1. Yellow body, black chassis, S-O-S Citroen decals, white Michelin figure on roof.
   2. Orange body and chassis, Garage du Pont decals.
   3. Light blue body, dark blue chassis, So.di.tec logo. Promo.

4411 CITROEN C4 AMBULANCE 106 mm 1/43 1984-1985
Same model as #4407, with white box and hubs, red Centrale Sanitaire Internationale decals.
   1. White body, chassis and box.

4412 DODGE 1940 PICKUP TRUCK 115 mm 1/43 1986-1988
Covered pickup with cast front and rear body, plastic windows and lights, white rear cover, tan interior, silver grille, bumpers and running boards, wheels with silver hubs, whitewalls, Flower Shop decals.
   1. Metallic blue body.

4413 DODGE 1940 PICKUP TRUCK 115 mm 1/43 1986-1987
Same basic model as #4412, but with flat white rear cover, Ship Chandlers decals on doors.
   1. Red-brown body.

4414 DODGE 1940 PICKUP TRUCK 113 mm 1/43 1986-1988
Same basic model as #4412, but with black rear sides-load of coal, J. Hansen Coal Co. decals.
   1. Metallic dark gray body.

4415 DODGE 1940 FIRE TANKER 112 mm 1/43 1986-1988
Tanker with cast body and chassis, plastic windows, red tank, silver caps, grille, bumper and running boards, wheels with silver hubs, whitewalls, Sellersville Fire Dept. decals.
   1. Red body and chassis.

4416 CITROEN C4 BP VAN 106 mm 1/43 1986-1987
Same model as #4407, but with green rear box, yellow BP Energol decals, light green tampo-printed stripes, yellow hubs.
   1. Yellow body, green chassis.

4417 AEC OPEN TOP DOUBLEDECK BUS 159 mm 1/50 1986-1987
Doubledecker with cast upper and lower body, plastic windows, blue interior and hubs, silver radiator shell, black grille and chassis, white divider and pillar, Union Jack and Tour of London decals.
   1. Blue upper and lower body.
   2. Light yellow and light green body, Vectis logo. Promo.

4418 DODGE 1940 MILK TANKER 112 mm 1/43 1986-1988
Same model as #4415, with L. F. Briggs Co. Milk decals.
   1. Blue body and chassis, white tank.

4419 DODGE 1940 TEXACO TANKER 112 mm 1/43 1986-1988
Same model as #4415, with Texaco Motor Oil decals.
   1. White body and chassis, red tank.

4420 DODGE 1940 FIRE PICKUP 115 mm 1/43 1986-1988
Same model as #4413, with Lampeter Fire Dept. decals.
   1. Red body, white rear cover.
   2. White body. Promo?

4421 DODGE 1940 COVERED PICKUP 115 mm 1/43 1987-1988
Same model as #4412, with Sun Club decals.
   1. White body, green cover.

2. White body, Paganetti logo. Promo.
3. Olive body, white cover, Argus de la Miniature logo. Promo.

4422   CITOREN C4 FIRE TANKER   101 mm   1/43   1987-1991

Tanker with cast front body, chassis, plastic windows, red tank with silver caps, silver radiator shell, lights and running boards, black grille, wheels with red hubs, spare wheel, Sapeurs Pompiers (one or two lines) decals.
  1. Red body, black chassis.

4423   DODGE 1940 OPEN TRUCK   114 mm   1/43   1987-1988

Truck with cast front body and chassis, plastic windows, white open rear body, tan interior, silver grile, bumper and running boards, wheels with silver hubs, whitewalls, blue and white dolphin and compass decals.
  1. Dark blue body and chassis.

4424   DODGE 1940 STP WRECKER   117 mm   1/43   1988-1989

Wrecker with cast front body, chassis, unpainted boom and mount, plastic windows, silver grille, bumper, running boards and rear bed, black interior and hook, wheels with silver hubs, spare wheel, R & T and STP decals.
  1. Yellow front body, black chassis.

4425   DODGE 1940 COVERED FIRE PICKUP   114 mm   1/43   1988-1989

Same basic model as #4412, but with black cover, silver ladder, Beverly Hills Fire Dept. and F.D. decals.
  1. Red body, black cover.

4426   DODGE 1940 TEXACO TANKER   112 mm   1/43   1988-1989

Same model as #4415, with Texaco emblem and lettering decals.
  1. Red body and chassis, white tank.

4427   DODGE 1940 PEPSI-COLA TRUCK   118 mm   1/43   1989-1991

Flatbed with tailboard, with cast front body and chassis, plastic windows, blue rear body, brown load and interior, silver grille, bumper and running boards, wheels with silver hubs, whitewalls, Pepsi-Cola tampo-print.
  1. White body, red chassis.

4428   DODGE 1940 FIRE WRECKER   117 mm   1/43   1989-1991

Same model as #4424, with Chicago Fire Dept. decals.
  1. Red body, black chassis.

4429   CITROEN C4 KODAK VAN   106 mm   1/43   1991-

Same basic model as #4407, but with light orange rear cover, matching hubs, Kodak decals.
  1. Light orange body, red-orange chassis.

4430   DODGE 1940 SUNLIGHT SOAP TRUCK   118 mm   1/43   1991-

Same model as #4427, with dark blue rear body, no load, Sunlight Soap decals.
  1. Yellow body and chassis.

4431   FORD V8 NEWSPAPER VAN   108 mm   1/43   1992-

Panel van with cast body, chassis, plastic windows, black grille, silver radiator shell, lights and bumpers-inner chassis, wheels with blue hubs, whitewalls, black and white New York Times logo.
  1. Blue body, black chassis.

4440   DODGE 1940 WRECKER

I am not sure where I found this number listed. It could be the Los Angeles fire wrecker, but I suspect it is not official.

4455   CHEVROLET US SHERIFF'S CAR   114 mm   1/43   1988?

Same basic model as #4510, but with white door panels with County Sheriff star emblems. Number may not be official.
  1. Black body, dark gray interior.

4456 CHEVROLET R.C.M.P. CAR 114 mm 1/43 1988?
Same basic model as #4455, but with crown emblems on white door panels.
Number may not be official.
  1. Dark blue body, dark gray interior.

4500 SERIES

In 1984 a series of postwar classics appeared. At first it consisted of pairs of very similar models, each issued in two colors, but there have been exceptions to this pattern. The #4524 Tucker, for example, exists in just one form, with one catalog number, but in at least seven colors, and color variations have been added to earlier models.

4500 CADILLAC ELDORADO CONVERTIBLE 129 mm 1/43 9/1984-
Open convertible with cast body, matching opening hood, black chassis, plastic windshield, folded top, white interior, gray motor, silver bumpers, wheels with spoked hubs, whitewalls, interior detail labels.
  1. Red body, ivory or gray top.
  2. Pink body, gray top.
  3. Light blue body, gray top.
  4. Metallic blue body, black top.
  5. Light purplish body, gray top.

4501 CADILLAC ELDORADO HARDTOP 129 mm 1/43 9/1984-1988
Same basic model as #4500, but with white plastic hardtop, tan interior.
  1. Red body.
  2. Pink body.
  3. Yellow body.
  4. Light blue body.

4502 MERCEDES-BENZ 300SL COUPE 105 mm 1/43 10/1984-
Sports coupe with cast body, matching chassis, opening hood and right door, plastic windows and lights, interior, black and silver motor and grille, silver bumpers and trim, wheels with silver hubs.
  1. Red body, tan interior.

2. Metallic gray body, black interior.
3. Metallic gray body, Fimm-Montelimar logo. Promo.

4503 MERCEDES-BENZ 3005L RACING 105 mm 1/43 10/1984-1986
Same model as #4502, with red interior, #5 and racing stripes.
1. Silver body, orange stripes.
2. Silver body, blue stripes.

4504 FORD THUNDERBIRD CONVERTIBLE 120 mm 1/43 6/1985-1987
Open convertible with cast body, matching opening hood, black chassis, plastic windshield, interior with matching motor, white folded top, silver grille and bumpers, wheels with silver hubs, whitewalls.
1. Off-white body, brown interior.
2. Cream body, brown interior.
3. Red body, brown interior.
4. Yellow body, brown interior.
5. Light green body, brown interior.
6. Dark green body, white interior.
7. Metallic blue-gray body, red interior.
8. Light tan body, brown interior.
9. Light brown body, brown interior.

4505 FORD THUNDERBIRD HARDTOP 120 mm 1/43 6/1985-
Same basic model as #4504, but with white or black hardtop.
1. Pale green body, white hardtop and interior.
2. Pale green body, black hardtop, brown interior.
3. Metallic blue-green body, white hardtop, brown interior.
4. Blue body, white hardtop, black interior.
5. Metallic blue body, white hardtop, brown interior.
6. Metallic lilac body, white hardtop, red interior.
7. Black body, white hardtop, red interior.

4506 FERRARI 250 GTO 1958 101 mm 1/43 9/1985-
Sports coupe with cast body, matching opening doors, black chassis, plastic windows and lights, tan interior, silver exhaust pipes and parts, wheels with wire hubs, Ferrari emblems.
1. Red body.
2. Yellow body.
3. Blue body.
5. Metallic dark blue body.
6. Light purple body.
7. Red body, Fanauto and #88 logo. Promo.
8. Yellow body, Hobby News logo. Promo.

4507 FERRARI 250 GTO RACING 101 mm 1/43 9/1985-1986
Same model as #4506, plus #3 decals.
1. Silver body.
2. Dark green body.

4508 CHEVROLET 1950 SEDAN 114 mm 1/43 5/1986-
Sedan with cast body, black chassis, plastic windows and lights, tan or brown interior, silver grille, bumpers and parts, wheels with silver hubs, whitewalls. Plus/minus windshield centerpost.
1. Red body, brown interior.
2. Maroon body, tan interior.
3. Metallic dark green body, tan interior.
4. Metallic blue body, tan interior.
5. Metallic purplish-blue body, tan interior.
6. Metallic brown body, tan interior.
7. Pale green and olive body, Rétromobile 1988 logo. Promo.
8. Pale and medium blue body, Rétromobile 1988 logo. Promo.

4509 CHEVROLET 1950 TAXI 114 mm 1/43 5/1986-1987>#4529
Same basic model as #4508, plus Taxi sign on roof, black Checker Cab logo tampo-printed on doors. Plus or minus centerpost.

1. Yellow body, brown interior.

4510 CHEVROLET 1950 POLICE CAR 114 mm 1/43 1987-1988
Same basic model as #4508, plus siren on roof, black and white South Bend Police shield decals on doors.
1. Black body, brown interior.

4511 BUICK SUPER CONVERTIBLE 118 mm 1/43 2/1987-1989
Closed convertible with cast body, black chassis, plastic windshield, interior, black or white top, silver trim, lights, grille and bumpers, wheels with silver hubs, whitewalls, port decals.
1. Red body, tan interior, white top.
2. Pale pink body, tan interior, gray or black top.
3. Pale green body, tan interior, black top.
4. Light olive body, tan interior, black top.
5. Metallic blue body, tan interior, black top.
6. Metallic dark blue body, tan interior, black top.
7. Black body, tan interior, white top.

4512 BUICK SUPER CONVERTIBLE 118 mm 1/43 2/1987-1989
Same basic model as #4511, but with plastic folded top.
1. Red body, white top, tan interior.
2. Pink body, white top, black interior.
3. Maroon body, white top, tan interior.
4. Pale blue body, cream top, blue interior.
5. Metallic dark blue body, black top, tan interior.
6. Black body, white top, tan interior.
7. Red body, white top, Fimm-Montelimar logo. Promo.

4513 CHRYSLER WINDSOR SIX 123 mm 1/43 3/1988-
Sedan with cast body, black chassis, plastic windows, brown interior, silver trim, lights, grille, bumpers and parts, wheels with silver hubs, whitewalls.

1. Metallic dark green body.
2. Metallic blue body.
3. Metallic brown body.
4. Yellow and green body, Retromobile logo. Promo.
5. Light gray and red body, Retromobile logo. Promo.

4514 CHRYSLER WINDSOR TAXI 123 mm 1/43 1988-1990
Same basic model as #4513, plus taxi sign on roof, black Taxi logo tampo-printed on doors.
1. Yellow body, brown interior.

4515 FACEL VEGA HARDTOP 99 mm 1/43 6/1988-
Hardtop with cast body, black chassis, plastic windows and lights, black hardtop, gray or brown interior, silver grille, bumpers and trim, wheels with silver hubs.
1. White body, brown interior.
2. Red body, gray interior.
3. Dark blue body, gray interior.

4516 FACEL VEGA CABRIOLET 99 mm 1/43 6/1988-
Same basic model as #4515, but with white or black plastic folded top, gray, brown or black interior.
1. Red body, black interior and top.
2. Metallic silver green body, white top, gray interior.
3. Metallic dark blue body, white top, gray interior.
4. Black body, white top, brown interior.
5. Off-white body, black top, red interior. Promo?

4517 FORD THUNDERBIRD GRAND SPORT 120 mm 1/43 1988-
Sports car with cast body, matching headrest panel and opening hood, black chassis, plastic windows and lights, white interior and motor, silver grille and bumpers, wheels with wire hubs, whitewalls.

1. Red body.
2. Pale blue body.

4518 CHEVROLET FIRE CHIEF 114 mm 1/43 1988-1990
Same basic model as #4508, plus silver siren on roof, Fire Dept. and badge decals on doors.
1. Red body, brown interior.

4519 CITROEN 15 CV 1952 110 mm 1/43 #4102>1989-1990
Same model as #4102, with red or gray interior, yellow or pale green hubs.
1. Maroon body, red interior, yellow hubs, 1938-1988 decals.
2. Black body, gray interior, pale green hubs.
3. Black body, Raid Franco-Acadien logo. Promo.
4. Black body, Club Solido logo. Promo.
5. Pink body, Elodie logo. Promo.
6. Maroon body, Cinquantenaire 15 CV logo. Promo.
7. Dark blue body, Cinquantenaire 15 CV logo. Promo.
8. Light gray body, Cinquantenaire 15 CV lofo. Promo.
9. Set of ten different colors. Promo.

4520 CADILLAC ELDORADO SEVILLE 129 mm 1/43 1989-
Same model as #4501, with white interior and hardtop.
1. Metallic silver blue body.
2. Black body.

4521 STUDEBAKER SILVER HAWK COUPE 119 mm 1/43 9/1989-
Coupe with cast body, black chassis, plastic windows, blue roof, cream interior, silver grille, lights and bumpers, wheels with spoked hubs, whitewalls. Roof has rear side window pillars.
1. Light blue body.

4522 STUDEBAKER SILVER HAWK HARDTOP 119 mm 1/43 9/1989-
Same basic model as #4521, with dark green plastic hardtop having only one pillar at rear of each side.
1. Pale green body, brown interior.

4523 BUICK SUPER HARDTOP 118 mm 1/43 1990-
Same basic model as #4511, with white plastic hardtop.
1. Metallic dark blue body, blue interior.

4524 TUCKER 1948 SEDAN 122 mm 1/43 2/90
Sedan with cast body, black chassis, plastic windows, brown interior, silverbumpers and parts, wheels with silver hubs, whitewalls.
1. Metallic dark gold body.
2. Metallic dark red body.
3. Metallic blue body.
4. Metallic dark blue body.
5. Pale lilac body.
6. Metallic brown body.
7. Black body.

4525 CHRYSLER WINDSOR FIRE CHIEF 123 mm 1/43 2/1991-
Same model as #4513, plus gold and black Schenectady Fire Dept. decals.
1. Red body, brown interior.

4526 CITROEN 15 CV MONTE CARLO 110 mm 1/43 3/1991-
Same basic model as #4519, plus black plastic roof rack with spare wheel, silver hubs.
1. Black body, gray interior.

4527 TRIUMPH TR3A (not issued?)

4528 TRIUMPH TR3A LE MANS (not issued?)

4529 CHEVROLET 1950 TAXI 114 mm 1/43 #4509>3/1991
Same model as #4509, but with same black Taxi tampo-print used on #4514.
   1. Yellow body, brown interior.

4530 CHRYSLER POLICE CAR 123 mm 1/43 1992-
Same model as #4513, plus black and white South Bend Police labels.
   1. Black body, brown interior.

4531 FORD GT40 1964 ___ mm 1/43 1992-

4532 FORD GT40 LE MANS 1964 ___ mm 1/43 1992-

LIMITED EDITIONS AND PROMOTIONAL MODELS

There have been numerous limited editions and promotional models based on the 4500 series models, or sets of them that combine 4400 and 4500 models. There must be many more such models than I can list here, but I can only list and describe those that I own or otherwise know to exist. Let's begin with the limited edition fire vehicles, and go on from there. The Coca-Cola and Budweiser models are listed by their 9600 and 9100 numbers later.

Fire Vehicles

CADILLAC V16 1931: Red body, black roof panel and chassis, tan interior, emblem decals with yellow Boston Fire Department lettering on doors. Based on #4038.

CITROEN 15 CV 1952: Red body and hubs, coat of arms on door. Based on #4102.

CITROEN C4 COVERED TRUCK: Red body and chassis, black cab roof and rear cover, Service d'Incendie et de Secours decals on cover. Based loosely on #4403.

CITROEN C4 FIRE TRUCK: Probably identical to #4403 except for different decals.

CITROEN C4 TANKER: Red body, chassis and tank, white Service Départemental d'Incendie & de Secours decals on tank, Sapeurs Pompiers du Var on door. Based on #4422.

CITROEN TRUCK: Red body and chassis, black rear box, open at back, with silver ladder on top, white Centre de Secours de Bellac decal on box, plus white Tél. 61 on door. Based on 4400 series Citroen trucks with rear cover of #4425.

CITROEN TRUCK: Exactly like model just above, but with white Sapeurs Pompiers logo on box and door.

CITROEN WRECKER: Red body and chassis, with black-outlined gold Ville de Versailles logo on doors. Based on #4410.

CHEVROLET 1950 SEDAN: Red body, silver siren, red-yellow-black Philadelphia Fire Dept. emblem on door. Based on #4518.

CHRYSLER WINDSOR SEDAN: Red body, six-pointed Fire Chief star decal on door. Based on #4525.

DODGE FLATBED TRUCK: Red body and chassis, black flatbed, F.D.N.Y. decal on door.

DODGE TANKER: Red body, chassis and tank, Schenectady N.Y. Fire Dept. decal on tank, Schenectady badge on door. Based on #4415.

DODGE WRECKER: Red body and chassis, Boston Fire Department emblem on door. Based on #4428.

DODGE WRECKER: Exactly like model just above, but with Los Angeles County Fire Dept. decal on door.

These models probably date from 1989 or 1990. There may be more of them, and the individual models and a boxed set of twelve may well have catalog numbers.

Other Promotional Vehicles

AEC DOUBLEDECK BUS: Cream body, Vectis logo. Based on #4417.

CITROEN 15 CV: Yellow-orange body, blue VII. ICCCR lettering on roof, emblem with map of Germany on door. Based on #4102.

CITROEN 15 CV: Blue body, white 8. ICCCR 1,2,3 September 1989, Flevohof, Flevoland, Holland lettering on roof, multicolored drawing based on map of Holland on door. Based on #4102.

CITROEN C4 VAN: Black body, chassis and box, gold Promotion Team AKAI Audio & Video logo on box and door. Based on #4407.

CITROEN C4 VAN: Blue body, chassis and box, gold British Meat, white Smithfield London on box, white British Meat emblem on door. Based on #4407.

CITROEN C4 VAN: Cream body, black chassis, white box, green square and black Burlington Air Express logo on box. Based on #4407.

CITROEN C4 VAN: Cream body, black chassis, white box, East Somerset Railway logo on box, No. 2 on door. Based on #4407.

CITROEN C4 VAN: White body and box, dark blue chassis, Expo '92 logo on box, figure on door. Based on #4407; 1992.

CITROEN C4 VAN: Dark green body, box, chassis and hubs, silver Galeries Lafayette logo. Based on #4407; 1984.

CITROEN C4 VAN: Cream body, black chassis, white box, red Humbrol lettering on box, red and black Airfix emblem on door. Based on #4407.

CITROEN C4 VAN: Yellow body, brown chassis, light tan box, Un Demi Siecle Solido 1932-1982, B. Azema, E.P.A. label on box, label showing Azema's two books on door. Based on #4407.

CITROEN C4 VAN: Metallic blue body, 26e Salon du Jouet logo. Based on #4407. 1987.

CITROEN C4 VAN: Red body, black chassis, Stop Pieces Hydrauliques de Freinage logo. Based on #4407. 1986.

CITROEN C4 VAN: Yellow body, black chassis, Cafe-Restaurant-Bocuse-Collonges logo. Based on #4407. 1986.

CITROEN C4 VAN: Light brown body, brown chassis, Menuiserie Joseph-Pierre et Fils logo. Based on #4407. 1986.

CITROEN C4 VAN: Maroon body, black chassis, Torpedo logo. Based on #4407. 1987.

CITROEN C4 VAN: Blue body, white cover, Airelec logo. Based on #4407. 1987.

CITROEN C4 VAN: Cream body, black chassis, Autoliners UK logo. Based on #4407. 1987.

CITROEN C4 VAN: Yellow body, blue chassis, Michelin logo. Based on #4407. 1987.

CITROEN C4 VAN: Light blue body, dark blue chassis, So.di.tec logo. Based on #4407. 1987.

CITROEN C4 VAN: Yellow body, green chassis, BP logo. Based on #4407. 1987.

CITROEN C4 VAN: Cream body, tan chassis, Rotary Club d'Issoire logo. Based on #4407. 1988.

CITROEN C4 VAN: Light green body, red chassis, white box, Fnath logo. Based on #4407. 1988.

CITROEN C4 VAN: Light orange body, purple chassis, L'est Republicain logo. Based on #4407. 1988.

CITROEN C4 VAN: Brown body, maroon chassis, Kodak logo. Based on #4407. 1988.

CITROEN C4 VAN: Gray body and chassis, Gardeil logo. Based on #4407. 1988.

CITROEN C4 VAN: White body, dark gray chassis, Gardeil-Mousse d'or logo. Based on #4407. 1988.

CITROEN C4 VAN: Dark blue body and chassis, Cercle Ferroviphile logo. Based on #4407. 1988.

CITROEN C4 OPEN TRUCK: Blue-black body and chassis, open rear with brown barrels, gold Joshua Tetley & Son Limited logo on rear body, Tetley's Fine Ales on door. Based on #4408.

DODGE COVERED PICKUP: Black body and chassis, white rear cover with black U.S. Marines Ambulance lettering, red crosses on top and door. Based on #4421.

DODGE WRECKER: Yellow body and chassis, unpainted boom and mount, red B E P and phone number decal on door. Based on #4424.

CLUB SOLIDO MODELS

CS 1   SPARK PLUG   76 mm   1987
Replica of original Bougie Gergovia model, with cast body, cast-in Club Solido lettering, wheels with silver hubs and whitewalls.
　1. Yellow body.

CS 2   DUESENBERG J   124 mm   1/43   1988
Open car with cast body, matching hood, chassis, plastic windshields, dark blue bodywork, tan seats, silver grille, lights, bumpers and spares, wheels with wire hubs and whitewalls.
　1. Blue body, dark blue chassis.

OTHER 4000 MODELS

4601   PRESTIGE DE SOLIDO SET   1/43   5/1980-1981
Boxed set of three Age d'Or models: 4046 Rolls-Royce, 4085 Cadillac and 4136 Bugatti. Standard colors.

5000 SERIES

Several truck tractors and semi-trailers have been listed separately.

5001   DAF F2800 tractor
5002   Mercedes 2624 tractor
5003   Magirus-Deutz tractor
5004   Saviem Renault H 875 tractor
5005   Mercedes 1217 K/32 tractor
5006   Iveco tractor
5007   Mack R600 tractor
5008   Mercedes tractor

5101   Tank semi-trailer
5102   Savoyarde semi-trailer
5103   Bulk Carrier semi-trailer
5104   Renault box semi-trailer
5105   Mack box semi-trailer
5106   Logger semi-trailer
5107   Motor Home semi-trailer
5108   A. R. E. semi-trailer
5109   Mack Fire Truck semi-trailer
5110   Zodiac semi-trailer
5111   Fire Pumper semi-trailer
5112   U.S. semi-trailer

KITS

These kits are made up of the components of Solido cars (all but one from the 1000 series, that one the 1181 Alpine, with the number 5 replacing the first digit of their regular numbers) and decals with which to convert them to models of cars that ran in actual races or rallies. There were usually three different sets of decals for each model. They came on the market with these numbers in 1980, most of them having existed in kit form previously with catalog numbers consisting of the ready-made model's number followed by the letter K, as indicated below, and were apparently withdrawn at the end of 1982.

5016   FERRARI DAYTONA   #16k>1980-1982
1. Red body, #56, Le Mans 1973.
2. Red body, #71, Le Mans 1974.
3. Yellow body, #36, Feraud, Le Mans 1972.

5018   PORSCHE 917-10   #18k>1980-1982
1. White body, #23, RC, Mosport 1973.
2. White body, #59, Atlanta 1973.
3. Black body, #13, Andial, Mosport 1976.

5019   VOLKSWAGEN GOLF   #19k>1980-1982
1. White body, #72, OYO, Monte Carlo Rally 1978.
2. Red body, #30, Monte Carlo Rally 1977.
3. Black body, #5, SEM, Monte Carlo Rally 1977.

5026   FORD CAPRI   #26k>1980-1982
1. Yellow and black body, #14, Gerstmann, Monza 1973.
2. Black and red body, #90, Le Mans 1974.
3. Black body, #95, Berthe Moline, Le Mans 1975.

5029   CITROEN CX   #29k>1980-1982
1. Red body, #98, Aseptogyl, Monte Carlo Rally 1978.
2. Metallic blue body, #18, Eurocasion, Morocco 1976.
3. Black body, #26, Portugal 1978.

5041   ALFA ROMEO 33   #41k>1980-1982
1. Red body, #2, WKRT, Monza 1975.
2. Red body, #2, Campari, Spa 1975.
3. Red body, #3, Redlefsen, Nürburgring 1975.

5044   FERRARI BB   6/1980-1982
1. ? body, #87, Chinetti, Le Mans 1978.
2. ? body, #88, Thomson, Le Mans 1978.

5050   PEUGEOT 504   #50k>1980-1982
1. White and red body, #17, Aseptogyl, Bandama 1975.
2. White and black body, #6, Peugeot, Morocco 1976.
3. Red and white body, #5, Gitanes, Bandama 1975.

5054   FIAT 131 ABARTH   #54k>1980-1982
1. White body, #3, Parmalat, Cremona 1977.
2. White body, #6, San Remo 1977.
3. White and black body, #13, Chequered Flag, RAC Rally 1977.

5058   RENAULT 5   #58k>1980-1982
1. Blue-white-red body, #140, Arnold, Gordini 1976.
2. Blue body, #138, Kindy, Gordini 1976.
3. Black body, #5, Lebeluga, Gordini 1976.

5068   PORSCHE 934   #68k>1980-1982
1. Red body, #59, Mecarillos, Le Mans 1977.
2. Black body, #61, TS, Le Mans 1976.
3. Black body, #65, Ecco, Le Mans 1978.

5069   ALFASUD TI   #69k>1980-1982
1. White and blue body, #15, Jousse.
2. Blue-white-red body, #24, Karland.
3. Black? body, #6, Danauto.

5070   OPEL GTE   #70k>1980-1982
1. White body, #26, Kleber, Mille Pistes 1978.
2. Yellow and green body, #26, Cévenole 1976.
3. Yellow and black body, #10, Euro Handler, Acropolis 1978.

5073   LANCIA STRATOS   #73k>1980-1982
1. White body, #1, Alitalia, Monte Carlo Rally 1977.
2. White and red body, #8, Bic, Acropolis Rally 1978.
3. White-black-red body, #5, Pirelli, Monte Carlo Rally 1978.

5075   BMW 3.0 CSI   #75k>1980-1982
1. White body, #1. Muir Miles, Paul Ricard 1970.
2. Yellow body, #15, Lumitas.

5081   PEUGEOT 104 ZS   #81k>1980-1982
1. White body, #5, Serre-Chevalier 1978.
2. Red body, #2, Serre-Chevalier 1978.
3. Red body, #70, Circuit de Glace, Monte Carlo Rally 1978.

5082   ALFETTA GTV   4/1980-1982
1. ? body, #7, Ile d'Elbe 1979.
2. ? body, #9, San Remo 1978.
3. ? body, #44, Monza 1978.

5087   ALPINE A442   (announced in 1980, not issued)

5089   BMW 530   4/1980-1982
1. Black body, #2, BP, Rouen 1978.

2. ? body, #7, Pau 1978.
3. ? body, #11, Dijon 1977.

5094 TOYOTA CELICA 6/1980-1982
1. White body, #18, Lombard, RAC Rally 1977.
2. Red and orange body, #62, Spa 1973.

5181 ALPINE 1800 #181k>1980-1982
1. White body, #73, Aseptogyl, Tour de France 1972.
2. Yellow body, #98, Dinin, Monte Carlo Rally.
3. Blue body, #95, Miko, Tour de France 1972.

5400 PAIRS

5401 BUGATTI ROYALE & ATALANTE 4/1980-1982

5402 ROLLS-ROYCE PHANTOM DROPHEAD & TOWN CAR 5/1980-1982

5403 DUESENBERG J LIMOUSINE & SPIDER 6/1980-1982

OTHER 5000

5998 DIORAMA SET 9/1980-1982
Set of four Peugeot 504 rally cars and one kit.

D-DAY MODELS

To commemorate the 40th anniversary of D-Day and the liberation of France, eight military models were reissued in 1984 and packaged either individually or in sets of four models.

601 D-DAY SET 1984
Set of M3 Halftrack, M10 Tank Destroyer, M20 Scout Car, Tiger Tank.

602 D-DAY SET 1984
Set of Dodge 6x6 Truck, Büssing 232, Panther Tank, Sherman Tank.

603 M10 TANK DESTROYER 135 mm #232>#2232>1984
Same model as #232. White US star decals.
   1. Olive body.

604 TIGER I TANK 164 mm #222>#2222>1984
Same model as #222. Red and white #131 decals.
   1. Dark gray body.

605 SHERMAN M4 TANK 124 mm #231>#2231>1984
Same model as #231. Yellow square decals.
   1. Olive body.

606 PANTHER TANK 196 mm #228>#2228>1984
Same model as #228. Black and white cross and #204 decals.

607 M20 COMBAT CAR 100 mm #200>#2200>1984
Same model as #200. White star decal.
   1. Olive body.

608   M3 HALFTRACK   127 mm   #244>#2244>1984
Same model as #244.
1. Olive body.

609   DODGE 6X6 AMBULANCE TRUCK 120 mm   #242>#2242>1984
Same model as #242.
1. Olive body, red cross decals.

610   BÜSSING 232 ARMORED CAR   114 mm   #226>#2226>1984
Same model as #226. Black and white cross and emblem decals.
1. Dark gray body.

OVERLORD SERIES

Five years later, in 1989, a new D-Day Series appeared for the 45th anniversary of the event. It was composed of twelve models, all using existing castings. The individual models all have the basic number 4489 followed by 1 to 12.

4489-1   JEEP AND TRAILER   140 mm   #256>#2256>1989>#6048
Same model as #256, with cast bodies, olive plastic chassis, Jeep top, folding windshield, parts and hubs, black wheels, white star decals.
1. Olive bodies.

4489-2   GMC "LE ROI" COMPRESSOR"   126 mm   #3116>1989>6001
Military version of #3116, with cast cab and chassis, olive plastic rear body parts, cab roof and hubs, black wheels, white star decals.
1. Olive cab and chassis.

4489-3   GMC OPEN CAB TRUCK WITH GUN   130 mm   1989>6047
Covered army truck with cast body and chassis, olive plastic cover, ring and hubs, black machine gun and wheels, white star decals.
1. Olive body and chassis.

4489-4   GMC COVERED TRUCK   130 mm   #3132>1989>6032
Military version of #3121, covered truck with cast front and rear body and chassis, plastic windows, olive cover and hubs, wheels, white star decals.
1. Olive body and chassis.

4489-5   DODGE AMBULANCE VAN   96 mm   #2128>1989>6043
Military version of #2128, van with cast body and chassis, plastic windows, olive opening rear doors, interior, inner chassis and hubs, wheels, red cross decals.
1. Olive body and chassis.

4489-6 DODGE SIGNAL CORPS VAN 96 mm #2128>1989>#6004

Same model as #4489-5, but with white star decals.
1. Olive body and chassis.

4489-7 DODGE 6X6 ARMY TRUCK 118 mm #242>#2242>1989>#6024

Same model as #242, with cast body and chassis, olive plastic cover and hubs, wheels, white star decals.
1. Olive body and chassis.

4489-8 M20 COMBAT CAR 98 mm #200>#2200>1989>#6028

Same model as #200, with cast body, chassis and ring, black plastic machine guns and parts, olive hubs, wheels, white star decals.
1. Olive body, chassis and ring.

4489-9 PACKARD HEADQUARTERS CAR 130 mm #4037>1989>#6006

Military version of #4037, with cast body and chassis, plastic windows, brown interior, olive grille and hubs, white bumpers, wheels, white star decals.
1. Olive body and chassis.

4489-10 M3 HALFTRACK 125 mm #241>#2241>1989>#6061

Same model as #241, with cast body, hood and chassis, olive plastic cover and running gear, front wheels, rear tracks, white star decals.
1. Olive body, hood and chassis.

4489-11 HANOMAG HALFTRACK 120 mm #241>#2241>1989

Same model as #241, with cast body and chassis, unpainted tracks, black plastic machine guns and front wheels, tan hubs and rear wheels, black and white cross decals.
1. Tan body and chassis.

4489-12 BÜSSING 232 SCOUT CAR 114 mm #226>#2226>1989>#6051

Same model as #226, with cast body, chassis and turning turret, black plastic antenna, wheels with tan hubs, black and white cross decals.
1. Tan body.

6000 SERIES

Azema lists a series of twelve military models issued in 1985, of which I know nothing. I can only list them here:

6001 Kaiser Jeep M34 Truck (same model as #244)
6002 Saviem V.A.B. Amphibian (same model as #251)
6003 Commando Police Amphibian (same model as #224B)
6004 Berliet Alvis (same model as #247)
6005 Patton M47 Tank (same model as #202)
6006 AMX 30 Tank (same model as #209)
6007 AMX 13 VTT Tank (same model as #227B)
6008 AMX 13/90 Tank (same model as #230)
6009 PT 76 Tank (same model as #207)
6010 SU 100 Tank (same model as #208)
6011 Leopard Tank (same model as #243)
6012 M7 B1 "Priest" (same model as #252)

LES MILITAIRES

Beginning in 1986, Solido issued a new series of military models, most of them revisions of earlier models, and the rest either based on the earlier military models or on existing civilian vehicles. One, in fact, combines the hood of the old Renault car with the body and chassis of the Citroen C4 truck. I had assumed that each was issued in just one form until I found the white version of the #6064 Jagdpanther, and then the white UN version of the #6027 VAB. There is at least one other.

6001 GMC 6X6 COMPRESSOR TRUCK 128 mm 1/1991-
Military version of #4411, with cast cab and chassis, olive plastic rear body parts, cab roof and hubs, white star decal.
   1. Olive cab and chassis.

6002 GMC 6X6 WRECKER 156 mm 2/1991-
Military version of #3117, with cast front and rear body and chassis, black plastic boom, olive hubs, white star decal.
   1. Olive body and chassis.

6003 CADILLAC 1930 STAFF CAR 126 mm 1/1991-
Military version of #4085, with cast body and chassis, olive plastic bumpers and hubs, black grille, white star decals.
   1. Olive body and chassis.

6004 DODGE SIGNAL CORPS VAN 91 mm 1992
Military version of #2128, with cast body and chassis, olive opening rear doors and hubs, white star and other decals.
   1. Olive body and chassis.

6005 KAISER JEEP COVERED TRUCK 140 mm 1992-
Based on #245, with cast front and rear body and chassis, tan plastic cover

and hubs, white star decal.
   1. Tan body and chassis.

6006 PACKARD HEADQUARTERS CAR 130 mm 1992-
Military version of #4037, with cast body and chassis, olive plastic bumpers and hubs, black grille, white star decals.
   1. Olive body and chassis.

6007 VAB 4X4 AMPHIBIAN 120 mm 1992-
Four-wheel version of #251, with cast body, turret and chassis, tan plastic parts and hubs.
   1. Tan body and chassis.

6020 CITROEN C4 ARMY AMBULANCE 101 mm 1986-1987
Military version of #4411, with cast body and chassis, olive plastic rear box, grille and hubs, red cross decals.
   1. Olive body and chassis.

6021 CITROEN C4 ARMY TRUCK mm 1986
MIlitary version of #4416, with cast body and chassis, olive plastic cover, grille and hubs.
   1. Olive body and chassis.

6022 UNIC TROOP TRANSPORT TRUCK 105 mm 1986
Based on #235, with cast body, cab and drive train, olive plastic hubs, seated soldiers.
   1. Olive body, cab and drive train.

6023 RENAULT KZ ARMY TRUCK 103 mm 1986-1987
Covered truck with cast hood, rear body and chassis, olive plastic cab roof, rear cover and hubs, drive figure.
   1. Olive hood, body and chassis.

6024 DODGE 6X6 ARMY TRUCK 118 mm #242>#2242>1986
Same model as #242, with olive plastic cover, red cross decals.
   1. Olive body and chassis.

6025 PANHARD AML 90 108 mm #240>#2240>1986-
Same model as #240, with cast body and turret.
   1. White body, black UN logo.
   2. Olive body and turret, Flandre and #5 decals.

6026 UXM 706 COMMANDO CAR 112 mm #224>#2224>1986-1987
Same model as #224, with cast body, turret and chassis, white decals.
   1. Olive body, turret and chassis.

6027 VAB 4X4 ARMORED CAR 120 mm 1986-1991
Same model as #6007, with matching plastic parts.
   1. White body and chassis, black UN logo.
   2. Olive body and chassis.

6028 M20 COMBAT CAR 100 mm #200>#2200>1986-1990
Same model as #200, with cast body and chassis, olive plastic parts, white star decal.
   1. Olive body and chassis.

6029 KAISER JEEP 6X6 TRUCK 140 mm #245>#2245>1986-1988
Same model as #245 and #6005, with plastic cover and hubs.
   1. Tan body and chassis.
   2. Olive body and chassis.

6030 RENAULT 4X4 COVERED TRUCK 96 mm #203>#2203>1986-1987
Same model as #203, with olive plastic covers, red extinguisher.

1. Olive body.
2. Olive body, red cross decals.

6031 BTR 40 ROCKET LAUNCHER 110 mm #225>#2225>1986
Same model as #225, with red rockets, red star and #416 decals.
1. Olive body, hatch and chassis.

6032 GMC ARMY TRUCK 128 mm 1987-1989
Based on 3100 series trucks, with cast open cab, rear body and chassis, olive plastic cab roof and rear cover, white star decal.
1. Olive cab, body and chassis.

6033 CHEVROLET 1950 STAFF CAR 116 mm 1987-1991
Military version of #4508, with cast body, black chassis, olive grille, bumpers and hubs, white star decals.
1. Olive body.

6034 WILLYS JEEP AND TRAILER 153 mm 1987-1990
Jeep from #236 with cast body, olive plastic top, pulling four-wheel trailer with cast body, olive plastic cover and chassis.
1. Olive Jeep and trailer bodies.
2. Olive bodies, 11e B.C. logo.

6035 CITROEN 15CV FFI CAR 110 mm #32B>#4039>1987-1988
Same model as #32B, with cast body, hoods, black chassis, olive plastic grille and bumpers, white FFI tampo-print.
1. Light and dark olive camouflage body and hoods.

6036 GMC FIVE-TON TRUCK 124 mm 1988-1989
Based on 3100 series trucks, with cast closed cab, rear body and chassis, olive plastic cover and grille, white star decal.
1. Olive cab, body and chassis.

6037 DKW MUNGA AND TRAILER 140 mm #212-213>#2213>1988-1990
Same model as #212-213, with cast bodies and DKW chassis, black plastic seats, grille guard and trailer chassis, olive hubs.
1. Tan bodies and chassis.

6038 UNIMOG RED CROSS TRUCK 106 mm 1988-1989
Based on 2100 series trucks, with cast cab and body, olive plastic cover, stack and hubs, red cross decals.
1. Olive cab and body.

6039 ARMY LAND-ROVER & TRAILER 165 mm 1988-1989
Same car as #66, same trailer as #213, with cast bodies and car chassis, olive plastic roof and hubs, black trailer chassis.
1. Olive bodies and chassis.

6040 DODGE PICKUP TRUCK 115 mm 1988
Military version of #4113, with cast body and chassis, olive plastic cover and bumper, white star decal.
1. Olive body and chassis.

6041 JEEP & RUBBER BOAT TRAILER 167 mm 1989-1991
Jeep from #256, with cast body, olive plastic cover, pulling olive plastic boat on olive trailer from #3000.
1. Olive body.

6042 CHRYSLER WINDSOR HEADQUARTERS CAR 124 mm 1989-1990
Military version of #4513, with cast body, black chassis, olive grille and bumpers, silver parts, white star decals.
1. Olive body.

6043 DODGE ARMY AMBULANCE 97 mm 1989-1990
Military version of #2128, with cast body, olive plastic chassis and bumper, white rear doors, red cross decals.
   1. Olive body.

6044 GMC ARMY FIRE TRUCK 128 mm 1990
Closed-cab version of ##6001, with cast cab and chassis, olive plastic rear body parts, white star decal.
   1. Olive cab and chassis.

6045 GMC ARMY WRECKER 155 mm 1990
Covered-cab version of #6002, with cast cab and chassis, olive plastic cab roof, black boom, white star decal.
   1. Olive cab and chassis.

6046 UNIMOG AMBULANCE 111 mm 1990-
Based on 2100 series trucks, with cast cab and chassis, olive plastic rear body, clear dome lights, red cross decals.
   1. Olive cab and chassis.

6047 GMC 6X6 ARMY TRUCK 132 mm 1990-
Same basic model as #6032 minus cab roof, with olive plastic ring and black machine gun, white star decal.
   1. Olive body and chassis.

6048 U.S. ARMY JEEP & TRAILER 139 mm 1991-
Same Jeep as #6041 with olive plastic top, white star decal, pulling same trailer as #6039 with black plastic chassis.
   1. Olive Jeep and trailer bodies.

6049 U.S. ARMY JEEP WITH MACHINE GUNS 83 mm 1991-
Same Jeep casting as #6048 minus windshield and top, with olive plastic

equipment, black machine guns.
   1. Olive body.

6050 DODGE SIGNAL CORPS VAN 91 mm 1991?
See #6004. Apparently not issued as #6050.

6051 BÜSSING ARMORED CAR 118 mm #226>#2226>1986
Same model as #226 minus antenna, with cast chassis, aprons, body, turret and gun, olive hubs.
   1. Tan and olive camouflage.

6052 M3 HALFTRACK 128 mm #244>#2244>1986-1988
Same model as #244 minus cover, plus black plastic machine gun.
   1. Olive body, hood and chassis.

6053 SHERMAN TANK 123 mm #231>#2231>1986-1991
Same model as #231, with olive plastic chassis, white star and #328 decals.
   1. Olive body, turret and chassis.

6054 AMX 10 P 115 mm #254>#2254>1986-1987
Same model as #254, with olive plastic chassis and parts, two soldiers, white decals.
   1. Olive body, mount and gun.

6055 LEOPARD TANK 182 mm #243>#2243>1986-
Same model as #243, with cast body, turret, gun and chassis.
   1. Olive body, turret, gun and chassis.

6056 VAB 6X6 AMPHIBIAN 123 mm #251>#2251>1986-1987
Same model as #251, with olive plastic parts and hubs.
   1. Olive body and chassis.

6057 AMX 13 VTT TANK 112 mm #227>#2227>1986
Same model as #227, with white Bretagne decals.
  1. Olive body, turret and chassis.

6058 AMX 13 TANK 133 mm #250>#2250>1986-1989
Same model as #250, with olive plastic parts, white decals.
  1. Olive body, turret and chassis.

6059 AMX 13 TWO-GUN TANK 108 mm 1986
Similar model to #249, with cast body, turret and chassis, blakc plastic
guns, olive parts, #19 decal.
  1. Olive body, turret and chassis.

6060 AMX 30 TANK 185 mm #209>#2209>1986-1991
Same model as #209, with olive plastic parts, black machine gun, soldier
figure, white Beauvais decals.
  1. Olive body, turret and chassis.

6061 HANOMAG HALFTRACK 119 mm #241>#2241>1986-1988
Same model as #241, with tan plastic parts, black machine guns, German
cross decals.
  1. Tan body and chassis.

6062 AMX 13 VCI TANK 112mm #227B>#2227>1986-1987
Same model as #2227B, with red cross decals.
  1. Olive body and chassis.

6063 TIGER TANK 163 mm #222>#2222>1987-
Same model as #222, with German cross and #141 decals.
  1. Dark gray body, turret, gun and chassis.

6064 JAGDPANTHER TANK 194 mm #228>#2228>1987-
Same model as #228, with black plastic machine gun and parts, cross and
#123 decals.
  1. White body, chassis and gun, black cross, red #123.
  2. Dark gray body, chassis and gun, red cross, black/white #123.

6065 PATTON TANK 160 mm #202S>#2214>1987-
Same model as #202S, with tan plastic hubs, Arabic decals.
  1. Tan body, turret and chassis.

6066 AMX 10 LANCE ROCKET LAUNCHER 115 mm 1988
Similar model to #6054, with cast body and turret, olive plastic chassis and
parts, soldier figure.
  1. Olive body and turret.

6067 GENERAL LEE TANK 119 mm #253>#2253>1988-
Same model as #253, with olive plastic chassis, white star and #328 decals.
  1. Olive body, turret and guns.

6068 M10 TANK DESTROYER 133 mm #232>#2232>1988-
Same model as #232, with black plastic machine gun, olive chassis.
  1. Olive body, turret and gun.

6069 HALFTRACK RECOVERY VEHICLE 144 mm 1989-1991
Same basic model as #6052, with same black plastic boom as #6002, other
black parts, white star decals.
  1. Olive body, hood and chassis.

6070 KAISER CRANE TRUCK 149 mm 1989-1990
Same basic model as #6029 minus cover, plus cast crane boom and mount, unpainted telescopic boom, olive plastic hook, white star decal.
  1. Olive body, chassis, mount and boom.

6071 GENERAL GRANT TANK 119 mm 1989-
Similar model to #6067, with different cast turret, olive plastic chassis, US flag decals.
  1. Olive body, turret and guns.

6072 RENAULT R35 TANK 95 mm #233>#2232>1990-1991
Same model as #233, with roundel decals.
  1. Olive and light brown camouflage.

6073 PANZER IV TANK 140 mm #237>#2237>1990
Same model as #237.
  1. Dark gray body, turret and chassis.

6074 SOMUA S35 TANK 104 mm #234>#2234>1991-
Same model as #234, with black plastic parts, roundel decals.
  1. Olive and light brown camouflage.

6075 PT 76 TANK 132 mm #207>#2207>1991-
Same model as #207, with red star decals.
  1. Olive body, turret, gun and chassis.

6076 AMX 10 WITH MACHINE GUN 115 mm #254>#2254>1992-
Same model as #254, with tan plastic chassis and parts.
  1. Tan body and turret.

6077 SHERMAN TANK WITH PLOW 139 mm 1992-
Same basic model as #231 plus tan plastic plow, mount and chassis, white star and square decals.
  1. Tan body, turret and gun.

6078 SHERMAN EGYPTIAN TANK 148 mm 1992-
Same basic model as #231, with tan plastic chassis.
  1. Tan body, turret and gun.

6079 AMX 30 B2 TANK 180 mm 1992-
Same basic model as #6060, with different turret, olive plastic parts.
  1. Olive body, turret, gun and chassis.

6300-6600 CIRCUS MODELS

These are the same models previously issued with 330, 621 and 660 numbers, plus one set that combines them. Only the numbers have been changed.

6330   CIRCUS PUBLICITY TRUCK   112 mm   #330>1980-1981
6331   CIRCUS COVERED TRUCK   123 mm   #331>1980
6332   CIRCUS CAGE SEMI   223 mm   #332>1980
6333   CIRCUS BOX OFFICE TRUCK   223 mm   #333>1980-1981
6335   CIRCUS ANIMAL SEMI   228 mm   #335>1980
6336   CIRCUS STAKESIDE SEMI   228 mm   #336>1980
6337   CIRCUS CARAVAN SEMI   228 mm   #337>1980-1981
6338   CIRCUS STAKESIDE TRAILER   194 mm   #338>1980
6621   CIRCUS LAND-ROVER AND TRAILER   248 mm   #621>1980-1981
6634   CIRCUS CRANE TRUCK   139 mm   #334>1980-1981
6660   CIRCUS TENT, TRUCKS AND FIGURES   #660>1980-1981
6661   CIRCUS TENT SET   #661>1980-1981
6662   CIRCUS DRESSAGE RING SET   #662>1980-1981
6663   CIRCUS TRUCKS AND FIGURES   #663>1980-1981
6664   CIRCUS TRUCKS SET   #664>1980-1981
6665   CIRCUS MENAGERIE SET   (not issued)
6666   CIRCUS SET   12/1980-1982
Boxed set including 6330, 6333, 6334, 6337, 6621, figures, etc. Standard colors.

GIFT SETS

There have been numerous gift sets, dealer lots and displays, and other aggregations in existence since 1980, bearing various 7000, 7100 and higher numbers. I hope this list of the 7000 gift sets is complete, but I have next to no data on the higher numbers.

7000   DISPLAY BOXES   1984-1988
Two 12-model display boxes.

7001   AGE D'OR TRUCKS   1983-1986
Five Citroen C4 trucks: Ambulance, coal, fire, Michelin, wrecker.

7002   FIRE TRUCKS   1984-1985
Four fire trucks: Saviem, Dodge, ladder truck, Renault van.

7003   FIVE CARS   1983-1990
Five 1300 series cars to 1988, then five 1200 series cars.

7004   CAR TRANSPORTER AND SIX CARS   1984-1988
Six 1300 series cars on ex-#321 transporter.

7005   EUROPEAN TRUCKS   1984-1985
Three semis: DAF tanker, Renault freighter, Iveco motor home.

7006   RENAULT CAR TRANSPORTER & TRAILER   1984-
Same transporter as in #7004, without cars.
1. Red cab and lower decks, yellow upper decks.
2. Red cab, upper and lower decks.
3. Black cab and lower decks, yellow upper decks.

7007   AGE D'OR CARS   1984-1986
Either French or American cars.

7008 PORSCHE SET 1984-1985
Five Porsches: 924, 928, 934, 935 and 936.

7009 CITROEN 15CV SET 1984
Four cars: Black 15cv, cream taxi, red Fire, black FFI.

7010 FIRE TRUCK SET 1985-1988
Five trucks: Berliet Camiva, ladder, snorkel, Jeep, Citroen C35.

7011 PUBLIC WORKS SET 1985-1986
Berliet truck with trailer and bodies.

7012 AGE D'OR SET 1985-1986
Four American cars.

7013 BERLIET LOW LOADER 1986-
Same basic model as #305, but without load.
   1. Red cab, semi and ramps.

7014 FIVE DODGE TRUCKS 1986-1987
Dodge pickup, covered pickup, coal truck and two tankers.

7015 MOUNTAIN SET 1986-
Unimog snowplow truck, snowmobile with plow, and helicopter.
   1. Red bodies.

7016 AMERICAN AGE D'OR CARS 1986-1988
Cadillac, Thunderbird, Chevrolet, Chevrolet police car.

7017 MILITARY SET 1986-1988
Low loader and missile-firing tank.

7018 24-CAR SALES DISPLAY BOX 1986-1990
With 1300 or 1200 series cars.

7019 INTERVENTION SET 1987-1988
Peugeot 205, Citroen C35 van and Citroen CX wagon as ambulances.
   1. White bodies.

7020 GENDARMERIE SET 1987-1988
Peugeot 205 police car, Peugeot J9 police bus and VAB 4x4 police vehicle.
   1. Blue bodies.

7021 AGE D'OR DODGES 1987-1988
Four Dodge trucks: pickup, covered pickup, coal truck and tanker.

7022 20 FIRE TRUCKS year?
Appears to be sales display of fire trucks.

7023 12 HI-FI VEHICLES year?
Appears to be sales display of Hi-Fi series.

7024 EUROP'ASSISTANCE SET 1988-1990
White and red Saviem wrecker, white Citroen C35 van and white Renault 4L.

7025 US MILITARY SET 1988-1991
Olive GMC army truck, Jeep, 105 mm gun and figures.

7026 CAR TRANSPORTER AND 6 CARS 1988-
New version of 7004, with 1200 or 1300 series cars.

7027 FIRE TRUCK SET 1988-
Four red trucks: Forest fire truck, Berliet ladder truck, Mercedes snorkel and open GMC.

7028 GENDARMERIE SET 1988-
Blue Gazelle copter, Peugeot 205 police car and Peugeot 504 wagon.

7029 AMERICAN AGE D'OR 1988-1989
New version of #7016, with Cadillac, Thunderbird, Buick and Chevrolet taxi.

7030 D-DAY SET 1984
Boxed set of all eight D-Day models.

7031 OPERATION MILITAIRE 85 1985
Two sets of four military models each.

7032 PRESTIGE SET 1984-1985
Six gold- or silver-plated models.

7033 TELE-UNION SET 1985
Citroen Visa in RTL, Monte Carlo, SSR and Canada versions, Iveco 4x4 truck, and Alouette copter.

7034 PRESENTOIR GAMME COMPLETE date?
I have no idea what this is.

7035 AEROPORTS DE PARIS 1988-1990
Red Berliet airport fire truck, white Peugeot ADP minibus, yellow Renault Flyco van, and white Alouette helicopter.

7036 BERLIET LOW LOADER 1990-
New version of #7013 low loader.

7037 CAR TRANSPORTER 1990-
New version of #7006 car transporter.

7038 DUO FRANCE 1989-
Renault 40CV and Renault 25.

7039 DUO FRANCE 1989-
Bugatti Atalante and Peugeot 205.

7040 DUO BRITAIN 1989-
Rolls-Royce drophead and Corniche.

7041 DUO GERMANY 1989-
Mercedes 540K and Porsche 928.

7042 DUO USA 1989-
Cadillac V16 landaulet and Chevrolet Camaro.

7043 DUO ITALY 1989-
Ferrari 250 GTO and BB.

7044 "AUCHAN" 1989
#4002, 4003 and 4088 Age d'Or models.

7045 "AUCHAN" 1989
#4516, 4519 and 4048 Age d'Or models.

7046 "FATHER'S DAY" 1991-
Three cars: Age d'Or, Sixties and Today.

7047 YEAR'S END SET 1991-
Delage, Duesenberg, and Dodge coal truck.

7151A AMERICAN VEHICLES
Twelve American fire vehicles?

7151F FRENCH VEHICLES
Twelve French and American fire vehicles?

7153 LE MANS RETROSPECTIVE SET
The first set, issued in 1990.

7155 COCA-COLA VEHICLES 1990?
Six vehicles with Coca-Cola logo.

7160 LE MANS RETROSPECTIVE SET
The second set, issued in 1992.

7167 COLLECTION OF SIX CARS

7400-7451 Display racks and boxes of various models: See Azema Volume I, p. 391 for more information.

7524 TUCKER TORPEDO
The same model in six different colors.

? TOUR AUTO 63-64 SET
Boxed set of two #4506 Ferrari GTO and plastic Tour Auto 63-64 sign. Ferraris are red with tan interior, and silver with black interior.

7600 SERIES

The following sets from the 600 series were renumbered in 1980. All were withdrawn at the end of that year.

7610 PEUGEOT 504 & STAR CARAVAN #610-1980

7613 RENAULT 12 POLICE CAR & MOTORCYCLE #613>1980

7614 CITROEN CX 7 CARAVAN #614>1980

7615 FIAT X1/9 & MOTORCYCLE TRAILER #615>1980

7616 PEUGEOT 504 BREAK & HORSE TRAILER #616>1980

7617 WINTER SPORTS SET #617>1980

7618 RACING SET #618>1980

7619 VACATION SET #619>1980

7620 AUTOROUTE WRECKER AND CAR #620>19

PRESTIGE SERIES

In 1987 Solido began to produce large-scale miniatures. I cannot describe them, as I do not collect them--even I have to stop somewhere! But I'll list them for you.

8001 BUGATTI ROYALE 1/21 1987-

8002 FORD PICKUP TRUCK 1/19 1987-1990
Ford logo.

8002 FORD FIRE PICKUP 1/19 1987-1989

8004A FORD COVERED PICKUP TRUCK 1/19 1987-1989
Pneus Dunlop logo.

8004B FORD COVERED PICKUP TRUCK 1.19 1987-1989
Dunlop Tires logo.

8004C FORD COVERED PICKUP TRUCK 1/19 1988-1989
Pepsi-Cola logo.

8005 FORD FIRE TANK TRUCK 1/19 1988-1991

8006 ROLLS-ROYCE SILVER CLOUD II 1/20 1988-

8007 BENTLEY S2 1/20 1988-

8008 FORD ROADSTER, top down 1/19 1988-

8009 FORD ROADSTER, top up 1/19 1988-

8010 FORD COVERED PICKUP TRUCK 1/19 1989-1991
Perrier logo.

8011 CADILLAC ELDORADO CONVERTIBLE, top down 1/21.5 1989-

8012 CADILLAC ELDORADO CONVERTIBLE, top up 1/21.5 1989-

8013 FORD PICKUP TRUCK 1/19 1990-1991
Pepsi-Cola logo.

8014 VOLKSWAGEN CABRIOLET, top down 1/17 1990-

8015 VOLKSWAGEN CABRIOLET, top up 1/17 1990-

8016 VOLKSWAGEN BEETLE 1/17 1991-

8017 FERRARI 365 CABRIOLET, top down 1/18 1991-

8018 FERRARI 365 CABRIOLET, top up 1/18 1991-

8019 FORD PICKUP TRUCK 1/19 1991-
Kodak logo.

8020 FORD PICKUP TRUCK 1/19 1991-
Sunlight Soap logo.

8021 MINI COOPER S 1964 1982-
Green body.

8022 MINI COOPER S 1964 1982-
Maroon body.

8023   MINI COOPER S 1964   1982-
Rally car.

8024   FORD PICKUP TRUCK   1992-
Michelin logo.

CUSTOM SERIES

8301   FORD ROADSTER   1/19   1991

8302   VOLKSWAGEN CABRIOLET   1/17   1991-

8303   FORD PICKUP TRUCK   1/19   1991-

8304   VOLKSWAGEN BEETLE   1/17   1991-

8305   MINI-COOPER S   1992-

ACTUA SERIES

8501 CITROEN XM V6 1/18 1989-

8502 PEUGEOT 605 SV 1/18 1989-

8503 CITROEN ZX PARIS-DAKAR 1/18 1991-

8504 RENAULT 1992 1/18 1992-

9147 VOLKSWAGEN GERMAN MAIL CAR 1/18 1992-

1/43 SCALE MODELS

The models listed below are limited-issue or promotional models whose numbers are known to me. They include the Budweiser and Coca-Cola models, plus one model used as a commemorative by Solido itself.

9180 CADILLAC BUDWEISER VAN 130 mm 1/43 1990
Same model as 4400 series Cadillac vans, with black roof panel, whitewall tires, yellow Anheuser-Busch St. Louis Budweiser lettering and multicolored letter A and eagle design.
  1. Maroon body, black chassis.

9181 DODGE FLATBED TRUCK 118 mm 1/43 1990
Same model as #4427, with brown load, red Budweiser, King of Beer lettering on roof, red and white Budweiser emblem on doors.
  1. White cab, red chassis, blue rear body.

9182 DODGE PICKUP TRUCK 114 mm 1/43 1990
Same model as #4413, with multicolored letter A and eagle design on doors, red-white-blue Budweiser label design on rear cover.
  1. Red cab, chassis and rear body, white cover.

9183 CITROEN BUDWEISER VAN 105 mm 1/43 1990
Same model as #4407, with yellow Budweiser lettering and multicolored letter A and eagle emblem on yellow-bordered maroon panel on rear body, same emblem and yellow Anheuser-Busch
Budweiser lettering on doors.
  1. Maroon body, black chassis, tan rear body, doors and hubs.

9189 CITROEN L'AGE D'OR VAN 105 mm 1/43 year?
Same basic model as #4407, with cast body and chassis, plastic rear box, maroon and black wheel hubs, L'Age d'Or par Solido labels on box, white

Club de l'Auto on front of box, white Club de l'Auto emblems on doors, plinth lettered "25 éme Paris Deauville".
1. Maroon body and chassis, lighter maroon box.

COCA-COLA SERIES

Solido has issued several sets of Coca-Cola models. I have not been able to find all the numbers, and there may have been more models. I can list the following:

#? FORD THUNDERBIRD GRAND SPORT 120 mm 1/43 198_
Same model as #4517, with white interior, black plastic bottle on strut, black design (woman drinking from bottle) on hood, white Drink Coca-Cola logo on sides and back.
1. Red body.

#? CHEVROLET 1950 SEDAN 115 mm 1/43 198_
Same model as #4508, but in #4518 box. Plastic Coca-Cola bottle on roof, black bottle decal on hood, white Drink Coca-Cola decals on doors, white Coca-Cola decals on trunk.
1. Red body.

#? MATRA RANCHO 98 mm 1/43 date?
Same model as #1062, with Coca-Cola logo.
1. Red body.

9501 FORD COVERED TRUCK 1/19 1990-
Same model as #8019, covered pickup with Coca-Cola logo on cover and doors.

9502 FORD ROADSTER 1/19 1990-
Same model as #8009, top-up convertible with Coca-Cola logo on hood and doors.

9503 VOLKSWAGEN 1/17 1992-
Same model as #8016, with yellow body, Coca-Cola logo on hood, doors and sides.

9504 VOLKSWAGEN 1/17 1992-
Same model as #9503, with red body, Coca-Cola logo (different from #9503) on hood and doors.

9601 CADILLAC 1930 VAN 130 mm 1/43 1990-
Same model as #4060, with black roof panel, logo of hand holding bottle, white Drink Coca-Cola in bottles lettering, red panel with yellow border, Advertising Car Coca-Cola Bottling Co. logo on doors.
   1. Metallic dark green body, black chassis.

9602 CHEVROLET 1950 SEDAN 115 mm 1/43 1990-1991
Same model as #4508, with white Drink Coca-Cola lettering on red disc on doors.
   1. Yellow body.

9603 DODGE COVERED PICKUP 115 mm 1/43 1990-1991
Same model as #4421, with brown plastic cover, white Drink Coca-Cola ice cold lettering on red panel with yellow border, plus red The Coca-Cola Bottling Co. lettering on doors.
   1. Yellow body and chassis.

9604 CHRYSLER WINDSOR SEDAN 123 mm 1/43 1990-1991
Same model as #4513, with white Drink, black-outlined Coca-Cola lettering on doors.
   1. Red body.

9605 DODGE FLAT TRUCK 115 mm 1/43 1990-1991
Same model as #4427, with brown rear bed-tailboard, red Drink Coca-Cola on roof, white-on-red Coca-Cola ovals on doors.
   1. Yellow body and chassis.

9606 CITROEN C4F TRUCK 1/43 1990-
Same model as #4407, with tan rear box, white Drink Coca-Cola on red panel with yellow and green borders, red Six-Box and other lettering below, figure of Six-Box to front, plus white Ask your dealer lettering on doors.
   1. Black body and chassis.

9607 CHEVROLET 1950 SEDAN 115 mm 1/43 1992-
Same model as #9602, with Coca-Cola bottle cap design on doors, black-outlined white Coca-Cola lettering on hood.
   1. Red body.

9608 CHRYSLER WINDSOR SEDAN 123 mm 1/43 1992-
Same model as #9604, with Coca-Cola bottle on doors, red Drink Coca-Cola lettering on front fenders.
   1. Yellow body.

9609 DODGE FLAT TRUCK 115 mm 1/43 1992-
Same model as #9605, with white Advertising Car, Coca Cola Bottling Co. on doors, white Drink and black-outlined white Coca-Cola lettering on roof.
   1. Red body and chassis.

9610 CADILLAC ELDORADO CONVERTIBLE 130 mm 1/43 1992-
Same model as #4500, with white interior, black folded top, black-outlined red Coca-Cola lettering on doors.
   1. White body and hood.

9701 DOUBLEDECK BUS 158 mm 1/50 1992-
Same model as 4402, with black chassis and seats, white divider and pillar, brown hubs, white-on-red Refresh yourself, Drink Coca-Cola decals.
   1. Red upper and lower body.

OTHER SOLIDO MODELS

In addition to the 1200 series of models made in Portugal, there have been other Solido models manufactured in other countries. The Dalia firm of Spain has made many pre-1957 and 100 series models, giving them its own numbers; all of them are listed in Azema Volume I, as are those made by Brosol of Brazil, Buby of Argentina, Tekno of Denmark and Louis Marx of the USA and Hong Kong. All of these except the few Tekno models used the original Solido catalog numbers. It is interesting to note, though, that a few special versions, from local taxis to police motor scooters, were added to the usual range by these firms.

Most of these models were rarely, if ever, seen in the USA. The most familiar models are the Ferrari 330 P3 and Porsche Carrera 6 made in Hong Kong by Louis Marx and once readily available in the United States. As Azema provides a full list of these models, I shall not copy that list here.

In closing, let me say that I believe Solido models richly deserve their popularity among collectors and should remain highly prized miniatures in the future. I hope Solido will continue to produce high-quality miniature vehicles for many more years. They have brightened our lives for more than sixty years, and despite present-day rumors of financial problems, I believe Bertrand Azema is right in anticipating that they will still be around to celebrate their 100th anniversary.

Appendix: SOLIDO Four-Digit Numbers

All models in 1980-1981 Catalog, plus others produced in 1980, with previous numbers and subsequent history:

1001 Renault 4 Mail Van, ex-42a, 1976, to 1981
1002 Renault 4 Fire Van, ex-42b, 1976, >1325, 1983 to 1987
1005 Citroen Visa, (new 1980) >1302, 1980 to 1987
1006 Fiat Ritmo, (new 1980) >1303, 1980 to 1987
1007 Citroen 2CV, (new 1980) >1301, 1980 to 1987
1010 Renault 5, ex-10, 1973, >1317, 1981 to 1983
1012 Peugeot 104, ex-12, 1973, to 1982
1016 Ferrari Daytona, ex-16, 1973, to 1981
1017 Ford Mirage, ex-17, 1973, 1980 only
1018 Porsche 917, ex-18, 1973, to 1981
1019 VW Golf, ex-19, 1973, >1314, 1981 to 1984, >1358, 1987 only
1020 Alpine A441, ex-20, 1973, 1980 only
1021 Matra Bagheera, ex-21, 1973, 1980 only
1022 Renault 12 Break, ex-22, 1974, to 1981
1025 BMW 3000 Rally, ex-25, 1974, to 1981
1026 Ford Capri Rally, ex-26, 1974, to 1981
1027 Lancia Stratos, ex-27, 1974, 1980 only

1028 BMW 2002 Turbo, ex-28, 1974, to 1981
1029 Citroen CX 2200, ex-29, 1974, to 1981
1030 Renault 30, ex-30, 1975, to 1981
1032 Porsche 935, new 1980, to 1982
1033 Fiat X1/9, ex-33, 1974, to 1981
1034 Land Rover Fire, new 1980, >2104, 1985 to 1987
1039 Simca 1308 GT, ex-39, 1976, to 1981
1040 Peugeot 604, ex-40, 1976, 1980 only
1041 Alfa Romeo 33TT, ex-41, 1976, 1980 only
1042 Renault 4 Van, ex-42c, 1976, 1980 only (reissued later?)
1043 Renault 14, ex-43, 1977, >1309, 1981 to 1985
1044 Ferrari BB, ex-44, 1976, to 1981
1045 Ford Escort, ex-45, 1976, >1315, 1982 to 1986
1047 Mercedes 280, ex-47, 1977, to 1981
1049 Porsche 928, ex-49, 1977, to 1982
1050 Peugeot 504 Rally, ex-50, 1977, to 1981
1051 Porsche 924 Turbo, new 1980, to 1982
1052 Lancia Beta Coupe, ex-52, 1977, to 1981
1053 Ford Fiesta, ex-53, 1977, >1313, 1981 to 1985
1054 Fiat 131 Abarth, ex-54, 1977, to 1982

1055 Peugeot 504 V6, new 1980, to 1982
1056 Citroen 2CV6, ex-56, 1977, to 1981
1057 Alpine A442 Turbo, ex-57, 1977, to 1981
1058 Renault 5 Coupe, ex-58, 1977, to 1981
1059 VW Scirocco, new 1980, to 1982
1060 Simca 1308 Taxi, ex-60, 1977, to 1981
1061 Ford Escort Rally, ex-61, 1977, to 1981
1063 Porsche 930, ex-63, 1978, 1980 only
1065 Citroen CX Break, ex-65, 1978, to 1981
1066 Land Rover, ex-66, 1978, to 1981
1068 Porsche 934 Turbo, ex-68, 1979, to 1982
1069 Alfasud TI, ex-69, 1978, to 1981
1070 Opel GTE Rally, ex-70, 1978, to 1982
1072 Citroen LN, ex-72, 1978, to 1981
1073 Lancia Stratos blue, ex-73, 1979, to 1982
1074 Lancia Stratos white, ex-73b, 1979, to 1982
1075 BMW 3.0 CSL, ex-75, 1978, to 1982
1076 Talbot-Simca Horizon, ex-76, 1978, to 1981
1081 Peugeot 104ZS, ex-81, 1978, >1316, 1981 to 1985
1082 Alfetta GTV, ex-82, 1979, to 1981
1086 Porsche 936, ex-86, 1979, to 1982
1087 Alpine A442, ex-87, 1979, to 1982
1089 BMW 530 Rally, ex-89, 1979, to 1982
1090 Peugeot 305, ex-90, 1979, >1320, 1982 to 1985
1091 Renault 18, ex-91, 1979, >1318, 1982 to 1987
1094 Toyota Celica, new 1980, to 1982
1096 Jaguar XJ12, new 1980, to 1982, > 1501, 1988 to date
1097 Porsche 934, ex-68, 1979, 1980 only
1123 Peugeot 504 Ambulance, ex-23am 1974, to 1981
1124 Peugeot 504 Gendarmerie, ex-23b, 1974, 1980 only
1125 Peugeot 504 Break, ex-23c, 1977, to 1981
1126 Peugeot 504 Fire, ex-23d, 1979, to 1981
1181 Alpine 1600, ex-181, 1970, to 1981

1193 Citroen GS, ex-193, 1972, to 1981
1198 Porsche 917K, ex-198, 1972, to 1981
2200 M-20 Combat Car, ex-200, 1961, 1980 only
2201 Missile Launcher, ex-201, 1961, 1980 only
2202 Patton M47 Tank, ex-202, 1962, to 1982
2203 Renault 4x4 Truck, ex-203, 1962, to 1982
2204 105 mm Cannon, ex-204, 1962, to 1982
2205 105 mm Cannon on Wheels, ex-205, 1962, to 1982
2206 250/O Howitzer, ex-206, 1962, 1980 only
2207 SU 100 Tank, tan, ex-208s, 1969, 1980 only
2208 SU 100 Tank, olive, ex-208, 1964, to 1981
2209 AMX 30 Tank, ex-209, 1965, to 1982
2210 AMX 30 AA Tank, ex-209b, 1965, to 1982
2211 Berliet Tank Transporter, ex-211, 1967, to 1982
2212 Berliet Transporter & AMX 30, ex-211b, 1967, to 1982
2213 Auto-Union Jeep & Trailer, ex-212-213, 1966, 1980 only
2214 Patton M47 Tank, tan, ex-202s, 1962, 1980 only
2218 PT76 Rocket Tank, ex-218, 1971, 1980 only
2222 Tiger Tank, gray, ex-222, 1970, 1980 only
2223 Tiger Tank, tan, ex-222b, 1972, 1980 only
2226 Bussing 232, ex-226, 1975, to 1982
2227 AMX 13 VTT, ex-227b, 1972, 1980 only
2228 Jagdpanther, gray, ex-228, 1971, 1980 only
2229 Jagdpanther, camouflage, ex-228b, 1971, 1980 only
2231 Sherman M4 Tank, ex-231, 1972, to 1981
2232 M10 Destroyer, ex-232, 1972, 1980 only
2233 Renault R35 Tank, ex-233, 1973, to 1981
2234 Somua S35 Tank, ex-234, 1873, 1980 only
2236 Panther G Tank, ex-236, 1973, 1980 only
2237 PZ 14 Tank, ex-237, 1974, to 1982
2238 AMX 30/Pluton Missile, ex-238, 1977, to 1982
2239 Panhard AML 90, ex-240s, 1973, 1980 only
2240 Panhard AML H90, ex-240, 1973, 1980 only

2241 Hanomag Halftrack, ex-241, 1974, 1980 only
2242 Dodge 6x6 Covered Truck, ex-242, 1975, to 1982
2243 Leopard Tank, ex-243, 1975, 1980 only
2244 M3 Halftrack, ex-244, 1976, to 1981
2245 Kaiser M34 Covered Truck, ex-245, 1975, to 1981
2247 Alvis Stalwart, ex-247, 1975, to 1981
2248 M41 Tank Chaser, ex-248, 1976, 1980 only
2249 AMX 13 2-Gun Tank, ex-249, 1975, 1980 only
2250 AMX 13 90 Tank, ex-250, 1975, 1980 only
2251 Saviem VAB Amphibian, ex-251, 1976, to 1982
2252 M7B1 Priest Tank, ex-252, 1976, to 1981
2253 General Lee Tank, ex-253, 1978, to 1982
2254 AMX 10 Tank, ex-254, 1978, 1980 only
2255 Berliet Airport Foam Truck, ex-255, 1979, to 1981
2256 Covered Jeep & Trailer, ex-256, 1979, to 1982
2257 Renault Fuel Tanker Semi, ex-257, 1979, to 1981
2259 Citroen C35 Ambulance, ex-259, 1978, to 1981
2262 Richier Crane Truck, ex-262, 1979, to 1981
3300 Saviem Wrecker, ex-366, 1977, to 1985
3301 Saviem Fire Wrecker, ex-366b, 1977, >2102, 1985 to 1987
3302 Saviem Police Wrecker, ex-366c, 1978, 1980 only
3303 Saviem Fire Truck, new 1980, >2001 1983 >2101 1985 to date
3304 Fire Jeep & Pump Trailer, new 1980, to 1982
3305 Low Loader with Pipes, ex-305b, 1977, to 1981
3306 Ship Dumper, new 1980, to 1982
3307 Iveco 4x4 Truck, new 1980, >3007, 1985 only
3321 Car Carrier & Trailer, ex-321, 1974, to 1981
3350 Berliet Camiva Fire Truck, ex-350, 1972, >2107, 1985 to 1987
3351 Berliet Airport Foam Truck, ex-351, 1973, to 1981
3352 Berliet Aerial Ladder Truck, ex-352, 1973, to 1981
3353 Richier Crane Truck, ex-353, 1973, >3102, 1985 to date
3354 Berliet 4x4 & Pump Trl., ex-354, 1973, >2103, 1985 to 1987
3355 Peugeot J7 Bus, ex-355, 1976, 1980 only

3356 Volvo Dump Truck, ex-356, 1974, >3103, 1985 only
3357 Unic Sahara Dumper, ex-357, 1974, 1980 only
3358 Mercedes Cherry Picker, ex-358, 1974, >3105, 1985 only
3359 Simca Snowplow Truck, ex-359, 1974, >2108, 1985 to 1987
3361 Mercedes Ladder Truck, ex-361, 1975, >3101, 1985 to 1986
3362 Hotchkiss Fire Truck, ex-362, 1975, >2000, 1983, >2100, 1985 to 1987
3363 Magirus/DAF Covered Semi, ex-363, 1975, >3501, 1985 to date
3364 Mercedes Bucket Truck, ex-364, 1975, to 1981
3365 IH Track Excavator, ex-365, 1976, 1980 only
3367 Volvo Front Loaderex-367, 1977, >3104, 1985 only
3368 Citroen C35 Fire Ambulance, ex-368, 1977, >2002, 1983, >2116 1986 to 1987
3369 DAF Shell Tanker Semi, ex-369, 1977, to 1985
3370 Saviem Container Semi, ex-370, 1977, >3502, 1985 to 1986
3371 Citroen C35 Fire & Lifeboat, ex-371, 1977, >3000, 1983, >3100, 1985 to 1986
3372 Peugeot J7 Police Bus, ex-372, 1977, >2003, 1983 to 1986
3373 Mercedes Livestock Truck, ex-373, 1978, to 1982
3374 Iveco Dump Truck, ex-374, 1978, >2009, 1974, >2109, 1985 to 1987
3375 Berliet Fire Truck, ex-375, 1978, >2008, 1984, >2106, 1985 to date
3376 Mercedes Bulk Tanker, ex-376, 1978, >3505, 1985 only
3378 Mercedes Excavator Truck, ex-378, 1978, to 1981
3379 Mercedes Garbage Truck, ex-379, 1978, to 1982
3380 Peugeot J7 Fire Ambulance, ex-380, 1978, 1980 only, >2115, 1986 to 1987
3384 Mercedes Covered Truck, ex-384, 1979, to 1981
3385 Mercedes Horse Van, ex-385, 1979, to 1982
3386 Mercedes Propane Tanker, ex-386, 1979, to 1982
3388 Mercedes Log Truck, ex-388, 1979, to 1988
3391 Dodge Fire Truck & Trl., ex-391, 1978, >3004, 1983 to 1985
3510 Renault Farm Tractor, ex-510, 1977, to 1981

3512 Tractor & Dumping Trailer, ex-512. 1977, to 1981
3514 Tractor & Tank Trailer, ex-514, 1977, to 1981
3515 Silage Trailer, ex-515, 1977, to 1981
3516 Sprayer Tank Trailer, ex-516, 1977, to 1981
3650 Ma Ferme Set, ex-650, 1978, to 1981
3810 Gazelle Europ'Assistance, ex-381a, 1979, to 1982
3811 Gazelle Gendarmerie, ex-381b, 1979, to 1982
3814 Alouette Securite, new 1980, to date
3815 Gazelle Army Copter, new 1980, to date
4001 Mercedes SSKL 1931, new 1980, to 1982
4031 Delage D8/120 1930, ex-31, 1974, to 1987
4032 Citroen 15CV 1939, ex-32, 1974, >4102, 1987, >4519, 1989 to date
4033 Citroen 15CV Fire Chief, ex-32a, 1974, to 1986
4034 Citroen 15CV FFI, ex-32b, 1974, to 1982
4035 Duesenberg Spider 1935, ex-35, 1975, to 1987
4046 Rolls-Royce Phantom 1939, ex-46, 1976, to 1987
4048 Delahaye Figoni 1937, ex-48, 1977, to date
4051 Delage Coupe D8/120 1938, ex-51, 1977, to 1986, also 1151
4055 Cord L29 1930, ex-55, 1977, to date
4059 Renault 40CV Berline, ex-59, 1977, to date, also 1159
4062 Hispano-Suiza Phaeton 1926, ex-62, 1977, to 1985
4067 Mercedes 540K, ex-67, 1978, to 1987
4071 Rolls-Royce Sedanca, ex-71, 1978, to date
4077 Rolls-Royce Drophead, ex-77, 1978, to date
4078 Delahaye Cabriolet, ex-78, 1978, to 1985
4080 Cord Coupe 1931, ex-80, 1979, to date
4085 Cadillac 1930, ex-85, 1979, to date
4088 Bugatti Atalante 1939, ex-88, 1979, to date
4097 Renault Reina Stella 1934, ex-97, 1979, to date
4132 Mercedes SS 1928, ex-132, 1964, to 1981
4136 Bugatti Royale 1930, ex-136, 1964, to 1981
4137 Mercedes SS Torpedo 1928, ex-137, 1965, to 1982
4140 Panhard Town Car 1925, ex-140, 1965, to 1981

4144 Voisin 17CV 1934, ex-144, 1966, to 1982
4145 Hispano-Suiza 1926, ex- 145, 1966, to 1985
4149 Renault 40CV 1925, ex-149, 1967, to 1986
4154 Fiat Limousine 1929, ex-154, 1967, 1985
4156 Duesenberg 1931, ex-156, 1969, to date
4401 Renault Paris Bus 1936, new 1980, to date
4402 London Doubledeck Bus, new 1980, to date
5016 Ferrari Daytona Kit, ex-16k, 1977, to 1982
5018 Porsche 917/10 Kit, ex-18k, 1977, to 1982
5019 VW Golf Kit, ex-19k, 1979, to 1982
5026 Ford Capri Kit, ex-26k, 1977, to 1982
5029 Citroen CX Kit, ex-29k, 1977, to 1982
5041 Alfa Romeo 33TT Kit, ex-41k, 1977, to 1982
5044 Ferrari BB Kit, new 1980, to 1982
5050 Peugeot 504 Kit, ex-50k, 1978, to 1982
5054 Fiat 131 Abarth Kit, ex-54k, 1978, to 1982
5058 Renault 5 Coupe Kit, ex-58k, 1977, to 1982
5068 Porsche 934 Kit, ex-68k, 1979, to 1982
5069 Alfasud TI Kit, ex-69k, 1978, to 1982
5070 Opel GTE Rally Kit, ex-70k, 1978, to 1982
5073 Lancia Stratos Kit, ex-73k, 1979, to 1982
5075 BMW 3.0 CSI Kit, ex-75k, 1978, to 1982
5081 Peugeot 104ZS Kit, ex-81k, 1979, to 1982
5082 Alfetta Rally Kit, new 1980, to 1982
5087 Alpine A442 Kit, announced 1980, never issued
5089 BMW 530 Kit, new 1980, to 1982
5094 Toyota Celica Kit, new 1980, to 1982
5181 Alpine 1600 Kit, ex-181k, 1979, to 1982
5401 Bugatti Royale & Atalante, new 1980, to 1982
5402 Rolls-Royce Phantom Drophead & Sedanca, new 1980, to 1982
5403 Duesenberg J Limousine & Spider, new 1980, to 1982
6330 Circus Publicity Van, ex-330, 1979, to 1981
6331 Covered Circus Truck, ex-331, 1979, 1980 only

6332 Circus Cage Semi, ex-332, 1979, 1980 only
6333 Circus Box Office Truck, ex-333, 1979, to 1981
6334 Circus Crane Truck, ex-334, 1979, to 1981
6335 Circus Livestock Semi, ex-335, 1979, 1980 only
6336 Circus Stake Truck, ex-336, 1979, 1980 only
6337 Circus Caravan Semi, ex-337, 1979, to 1981
6338 Circus Stakeside Trailer, ex-338, 1979, to 1981
6621 Circus Land Rover & Trl., ex-621, 1979, to 1981
6660 Circus Vehicle Set, ex-660, 1978, to 1981
6661 Circus Transport Set, ex-661. 1979, to 1981
6662 Circus Dressage Set, ex-662, 1979, to 1991
6663 Circus Caravan Set, ex-663, 1979, to 1981
6664 Circus Big Top Set, ex-664, 1979, to 1981

Note: The numbers listed above are meant to represent only successive appearances of the same model. They are not meant to include every use of the basic casting. And they do not include a few 1100 series models that were renumbered very briefly in 1980 but not catalogued as such. In addition, a few models were announced with old numbers but not issued until the new numbers had been introduced in the spring of 1980.

The following Solido models have been announced for late 1992 and 1993. It is not known whether they have been or will be issued.
1527 Lamborghini Diablo 1990
1528 Renault Twingo 1992 with roof panel
1529 BMW Series 3 Capriolet 1993
1530 Renault Twingo 1992
1531 Unidentifiedin 1993 catalog
1813 Chevrolet Corvette 1968
1814 Alpine A310 1972
1815 Chevrolet Camaro Z28 1983

The new 1900 series is composed of racing vehicles, some of which are renumbered, others new issues.
1901 Mercedes SSKL 1930, ex-4004
1902 Farrari 250 GTO 1963, ex-4507
1903 Citroen 15CV Monte Carlo 1952, ex-4526
1904 Alpine A110 Monte Carlo 1973, (=1803)
1905 Renault Clio 168 Coupe, 1991, ex-1520
1906 Lancia Stratos 1978, ex-1073
1907 Ferrari 365 GTB4 1972 (=2410)
1908 Renault 5 Maxi Turbo 1985, ex-1353
1909 AC Cobra 427 1965
1910 Chevrolet Corvette 1968 (=1813)
1911 Iveco Transafrica 1980, ex-3007
1912 Mercedes Unimog Rallye 1992
4160 Alfa Romeo 2500 Sport cabrio 1939
4161 Alfa Romeo 2500 Sport coupe 1939
4432 Ford V8 Fire Dept. Wrecker 1936
4433 Ford V8 Covered Pickup 1936, Kodak logo
4533 AC Cobra 427 1965 (=1909)
4534 VW Kombi Bus 1966
4535 VW Kombi Fire Van, 1966
4536 Citroen 15 CV Taxi 1952, ex-4041
8025 VW Beetle 1958, 1/18 scale
8026 Ford Pickup Fire Truck 1936m 1/18 scale
8306 VW Beetle 1958 (custom version of 8025)
9611 Ford V8 Van 1936, Coca-Cola logo

The ex- or = means that the two are the same basic vehicle, but not necessarily the same in every detail.

PRICE GUIDE

Major Series: $100 and up
37-42 Guns: $20-30
43-43a-43b Military: $75-100

Junior Series: $60-80, except
70 Cabriolet: $80-100
71 Graham-Paige: $80-100
82 Ferrari: $80-100
84 Studebaker: $90-110
89 Nash: $125-150
90 Cadillac: $90-110
95 Fire Truck: $90-110
96 Ladder Truck: $90-110
114 Packard: $90-110
115 Studebaker: $250-300
191 Mercury: $250-300
195 Ford Thunderbird: $170-180
196 Ford Wagon: $135-150
196 Ford Ambulance: $100-120
197 Pickup Truck: $80-100

Other trucks: $40-60
Scooters, trailers: $35-50

Baby Series: $45-60, except
134 Bus: $60-75
135a Studebaker: $100-120
137 Nash: $90-110
140 Henry J: $140-160

Mosquito Series: $30-40

Aircraft: $50-75, except
175 Constellation: $75-100

100 Jaguar D: $75-95
101 Porsche Spyder: $80-100
102 Maserati F.1: $65-75
103 Ferrari 500: $80-100
104 Vanwall F. 1: $70-85
105 Mercedes 190SL: $75-90

106 Alfa Romeo Giulietta: $75-90
107 Aston Martin DBR1: $70-85
108 Peugeot 403: $50-65
109 Renault Floride: $50-65
110 Simca Oceane: $65-80
111 Aston Martin DB4: $75-90
112 DB Panhard: $65-75
113 Fiat Abarth: $70-85
113B Ford Thunderbird: $175-200
114 Citroen Ami 6: $55-65
115 Rolls-Royce: $115-130
116 Cooper F. 2: $65-75
117 Porsche F. 2: $70-85
118 Lotus F. 1: $65-75
119 Chausson Bus: $150-175
120 Chausson Trolley: $160-175
121 Lancia Flaminia: $60-75
122 Ferrari F. 1: $70-85

123 Ferrari 250GT: $75-95
124 Abarth 1000: $50-60
125 Alfa Romeo 2600: $50-65
126 Mercedes 220SE: $60-75
127 NSU Prinz IV: $50-65
128 Ford Thunderbird: $165-195
129 Ferrari 2.5: $70-85
130 Aston Martin Vantage: $60-75
131 BRM F. 1: $65-75
132 Mercedes SS top up: $50-65
133 Fiat 2300S: $60-75
134 Porsche Le Mans: $70-85
135 Lola-Climax F. 1: $65-75
136 Bugatti Royale: $60-75 black: $25-35
137 Mercedes SS top down: $30-40
138 Harvey Special: $75-85
139 Maserati 3.5: $60-75

140 Panhard-Levassor 1925: $40-50

141 Citroen Ami 6 Break: $50-65

142 Alpine F. 3: $50-60

143 Panhard 24BT: $40-50

144 Voisin 1934: $40-50

145 Hispano-Suiza 1926: $40-50

146 Ford GT40: $65-75

147 Ford Mustang: $180-195

147B Ford Mustang Rally: $180-195

148 Alfa Romeo GTZ: $50-65

149 Renault 40CV 1926: $30-40

150 Oldsmobile Toronado: $150-175

151 Porsche Carrera 6: $60-70

152 Ferrari 330 P3: $60-70

153 Chapparal 2D: $85-95

154 Fiat 525N 1926: $50-65

156 Duesenberg J 1931: $50-75

157 BMW 2000 CS: $65-80

157B BMW 2000 CS Rally: $70-85

158 Alfa Romeo & Caravan: $100-125

159 Citroen Ami 6 & Boat: $80-95

161 Lamborghini Miura: $65-75

164 Simca 1100: $50-60

164B Simca Police: $100-120

165 Ferrari GTB4: $60-75

166 De Tomaso Mangusta: $50-65

167 Ferrari F. 1: $60-75

167B Ferrari F. 1: $65-75

168 Alpine Renault: $50-60

169 Chapparal 2F: $75-85

170 Ford Mark IV: $50-60

171 Opel GT 1900: $50-60

172 Alfa Carabo Bertone: $40-50

173 Matra F. 1: $55-65

174 Porsche 908: $60-75

175 Lola T70: $55-65

176 McLaren M8B: $60-70

177 Ferrari 312P: $60-70

178 Matra 650: $45-55

179 Porsche 914/6: $40-50

180 Mercedes C111: $40-50

181 Alpine Renault: $35-45

182 Ferrari 512S: $50-60

183 Alfa Romeo Zagato: $35-45

184 Citroen SM: $40-50

185 Maserati Indy: $45-55

186 Porsche 917: $45-60

186M Porsche Martini: $50-60

187 Alfa Romeo 33/3: $45-55

188 Opel Manta: $40-50

188R Opel Manta Rally: $45-55

189 Bertone Buggy: $30-40

190 Ford Capri: $35-45

192 Alpine A310: $35-45

192B Alpine Gendarmerie: $35-45

193 Citroen GS: $35-45

194 Ferrari 312PB: $40-50

195 Ligier JS3: $40-50

196 Renault 17TS: $35-45

197 Ferrari 512M Sunoco: $40-50

197B Ferrari 512M Piper: $40-50

198 Porsche 917: $40-50

199 March Can-Am: $35-45

200 M-20 Combat Car: $60-75

201 Unic Rocket Launcher: $125-150

202 Patton Tank: $100-120

203 Renault Tous Terrains: $$75-95

204 105 F Cannon: $50-65

205 105 C Cannon: $60-75

206 250/0 Howitzer: $60-75

207 PT 76 Amphibian Tank: $100-120

208 SU 100 Tank: $100-120

209 AMX 30 Tank: $100-120

210 AMX Two-gun Tank: $100-120

211 Berliet Tank Transporter: $80-100

211B Transporter with Tank: $175-200

212-213 AU Jeep & Trailer: $50-65

214 Berliet Aurochs: $75-95

215 Renault Police Truck: $75-100

218 PT 76 Rocket Tank: $100-120

219 M 41 Tank Destroyer: $100-120

221 Alfa Romeo Police: $60-75

222 Tiger I Tank: $100-120

223 AMX A-A Tank: $100-120

224 XM 706 Commando: $80-100

225 BTR 40 Rocket Launcher: $80-100

226 Büssing Scout Car: $100-120

227 AMX 13 VCI: $90-110

228 Jagdpanther Tank: $100-120

229M Jagdpanther remote: $125-145

230 AMX 13 Tank: $100-120

231 Sherman Tank: $100-120

232 M10 Tank Destroyer: $100-120

233 Renault R35 Tank: $100-120

234 Somua S35 Tank: $100-120

235 Simca-Unic SUMB Tank: $100-120

236 Panther Tank: $100-120

237 Panzer IV Tank: $100-120

238 AMX 30 Pluton: $100-120

240 Panhard AML: $90-110

241 Hanomag Halftrack: $100-125

242 Dodge 6x6 Truck: $80-100

243 Leopard Tank: $175-195

244 T34/85 Tank: $160-180

245 Kaiser-Jeep 6x6 Truck: $80-100

247 Berliet Alvis: $70-85

248 M41 Tank Destroyer: $100-120

249 AMX 13 Two-gun Tank: $100-120

250 AMX 13 90 Tank: $100-120

251 Saviem VAB 6x6: $100-120

252 M7 B1 Priest: $100-120

253 General Lee Tank: $100-125

254 AMX 10P Amphibian: $100-120

255 Berliet Foam Tanker: $85-100

256 Jeep and Trailer: $60-75

257 Saviem Tanker Semi: $70-85

259 Citroen Army Ambulance: $65-80

262 Richier Crane: $75-90

300 Berliet Titan: $150-175

301 Unic Sahara with Tanks: $150-175

302 Willeme Horizon: $120-140

303 Berliet Dumper: $80-100

304 Bernard Refrigerator: $100-120

305 Berliet Low Loader: $125-150 with house: $200-225

305B Berliet with Pipes: $125-150

306 Berliet Stradair Dumper: $70-90

307 Berliet Stradair Covered: $50-65

308 Willeme Elf Tanker: $65-75

316 Saviem Open Semi: $100-120

317 Berliet Yoplait: $125-150

318 Saviem Elf Tanker: $50-65

319 Saviem Esso Tanker: $50-65

320 Saviem Shell Tanker: $50-65

321 Saviem Car Carrier: $75-95

330 Citroen Circus Van: $70-80

331 Mercedes Circus Truck: $75-85

332 Saviem Circus Cage Semi: $80-90

333 DAF Circus Ticket Office: $75-85

334 Richier Circus Crane: $75-85

335 DAF Circus Animal Truck: $80-90

336 DAF Circus Stake Truck: $75-85

337 DAF Circus Caravan: $70-80

338 Circus Stake Trailer: $60-70

350 Berliet Camiva Fire Truck: $35-45

351 Berliet Foam Tanker: $35-45

352 Berliet Ladder Truck: $35-45

353 Richier Crane Truck: $35-45

354 Berliet Forest Fire Truck: $35-45

355 Peugeot J7 Minibus: $40-50

356 Volvo-BM Dumper: $40-50

357 Unic Sahara Dumper: $40-50

358 Mercedes Overhead Service: $45-60

359 Simca-Unic Snowplow: $35-45

361 Mercedes Ladder Truck: $35-45

362 Hotchkiss Fire Truck: $30-40

363 Magirus Covered Semi: $40-50

364 Mercedes Bucket Truck: $35-45

365 International Shovel: $150-175

366 Saviem Wrecker: $30-45

367 Volvo BM Loader: $40-50

368 Citroen Fire Ambulance: $25-35

369 DAF Tanker Semi: $40-50

370 Saviem Cargo Semi: $40-50

371 Citroen & Lifeboat: $35-45

372 Peugeot Police Bus: $30-40

373 Mercedes Livestock Truck: $35-45

374 Iveco Dump Truck: $35-45

375 Berliet Fire Truck: $30-40

376 Mercedes Bulk Semi: $40-50

378 Mercedes Excavator: $45-55

379 Mercedes Garbage Truck: $35-45

380 Peugeot Fire Ambulance: $30-40

381 Gazelle Helicopter: $20-25

384 Mercedes Covered Truck: $30-40

385 Saviem Horse Van: $40-50

386 Mercedes Propane Tanker: $30-40

388 Saviem Log Truck: $35-45

389 Log Trailer: $20-30

391 Dodge Fire Truck: $25-35

510 Renault Tractor: $30-40

511 Tipping Trailer: $25-35

512 Tracotr & Trailer: $50-60

513 Tank Trailer: $25-35

514 Tractor & Tank Trailer: $50-60

515 Silage Trailer: $20-25

516 Sprayer Trailer: $25-35

600 Peugeot 504: $40-50

601 Opel Commodore: $40-50

602 Renault 16: $40-50

610 Peugeot & Caravan: $55-65

611 Opel & Boat Trailer: $50-60

612 Renault & Amimal Trailer: $55-65

613 Gendarmerie Set: $35-45

614 Citroen & Caravan: $35-45

615 Fiat X1/9 & Cycle Trailer: $25-35

616 Peugeot & Horse Trailer: $40-50

617 Winter Sports Set: $35-45
618 Racing Set: $30-40
619 Vacation Set: $35-45
620 Autoroute Set: $35-45
621 Circus Car & Trailer: $75-85
650 Farm Set: $100-125
660-664 Circus Sets: $150 and up

10-96 modern cars: $20-30 each
31 Delage D8 1939: $25-35
32 Citroen 15 CV: $20-35
35 Duesenberg Spider: $30-45
42 Promotional vans: $40-80
46 Rolls-Royce 1939: $30-40
48 Delahaye 1937: $30-40
51 Delage D8 1939: $25-35
55 Cord L29 1930: $25-35
62 Hispano-Suiza 1926: $30-40
67 Mercedes 540K 1939: $25-35
71 Rolls-Royce 1939: $25-35
77 Rolls-Royce 1939: $25-35
78 Delahaye 135M: $25-35
80 Cord L29 Coupe: $25-35
85 Cadillac V16 1930: $25-35
88 Bugatti Atalante: $25-35
97 Renault Reinastella: $25-35

Kits 10 or 100 series: $15-25 each
1000 series cars: $20-30 each
1100 modern cars: $20-30 each
1100 Age d'Or: $20-30 each

1200 series Toys: $15-25 each
1300 Cougar series: $20-30 each
1300 Solido series: $20-30 each
1500 Hi-Fi/To-Day: $15-25 each
1700 Solido 2: $20-30 each.
1800 series: current
2000 series trucks: $20-30 each
2100 series trucks: $20-30 each or current
2200 military: same as 200 series
3000 series: $20-35 each
3100 series: $20-35 each or current
3300 series: $20-35 each
3500 farm: same as 500 series
3500 semi series: $25-40 each or current
3600 snow vehicles: $20-25 each or current
3800 helicopters: $15-25 each or current
4000 Age d'Or: $20-35 each or current
4100 Age d'Or: $20-35 each or current
4400 Age d'Or: $20-30 each or current
4500 Age d'Or: $20-30 each or current
5000 series kits: $15-25 each
6000 military: $20-35 each or current

D-Day military: $35-45 each
7000 gift sets: current or slightly over retail
8000 1/18 scale: current or slightly over retail
9000 promotional: current or up to twice retail

Note: "Current" refers to what was in production and on the market in 1992 or early 1993.

Many thanks to James Wieland for his help with the price guide.

Additions, corrections and comments are welcome; please send to:

Dr. Edward Force,
42 Warham Street,
Windsor CT 06095 USA.